GOOD HOUSEKEEPING FREEZER HANDBOOK

GOOD HOUSEKEEPING FREEZER HANDBOOK

Compiled by the
Good Housekeeping Institute

EBURY PRESS
LONDON

Published by Ebury Press
Division of The National Magazine Company Ltd.,
Colquhoun House
27–37 Broadwick Street
London W1V 1FR

First published in 1975
Revised edition 1983
Second impression 1985
Third impression 1985
Fourth impression 1987
Fifth impression 1988
Sixth impression 1989

ISBN 0 85223 283 7 (paperback)
ISBN 0 85223 294 2 (hardback)

Illustrated by Ivan Ripley
Photographs by Paul Kemp

The publishers would like to thank Divertimenti for their help in lending props for photography

Filmset by Advanced Filmsetters (Glasgow) Limited
Printed and bound in Great Britain at The University Press, Cambridge

CONTENTS

INTRODUCTION

Frozen food is one of the miracles of modern life. Look back to the days when preserving food involved lengthy curing, smoking, pickling and bottling—all masking the original fresh flavour—and think how much easier and quicker it is to put your prepared food into a machine which then does all the preserving work for you. Provided you freeze the right things, frozen food retains its colour, texture, taste *and* nutrients. Owning a freezer means you need shop less often, cook only when you feel like it, easily preserve produce straight from the garden and any you might buy cheap at the height of its season. And you'll always have meals at the ready for unexpected occasions.

Why buy a freezer?

The freezer itself will cost money to buy and money to run. To get the best from it you must be prepared to shop and cook to ensure that it stays full and works to best advantage. That said, freezers are a real boon for most people and a virtual necessity for some.

People who are out working, or busy looking after children will find a freezer a real boon. There will be no need to spend frantic lunch hours shopping and then struggling home with bursting carrier bags. Go on a major shopping expedition once a fortnight, or even less often—you'll save precious time and have food stored away to see you through any emergencies.

Keen gardeners, you can freeze your surplus produce and have fresh vegetables and fruit—possibly even be self-sufficient—all year round. There is no need to get tired of seasonal gluts, and you'll have the opportunity to experiment with new varieties specially developed for freezer owners.

For *elderly people* there's no need to shop when it rains or you're tired and you can buy ready-prepared frozen meals you'd otherwise have had to cook from scratch. If you live alone you can freeze food in the size portions you want to eat.

If you live in a rural area it's no fun being snowed up for weeks on end—even less so if your diet consists entirely of tinned food. And with a freezer you can save transport costs by making fewer shopping trips.

People who entertain frequently may get bored with boeuf bourguignonne four weeks running but your guests won't. And think of the time and fuel you'll save by cooking the whole lot in one go and freezing it in dinner party-sized packs.
Unfortunately, *people who love cooking when the mood takes them* often find this isn't usually when the rest of the household wants a meal. With a freezer you can succumb to the desire to cook whenever you want... and there will be meals ready when everyone else wants to eat.

Twenty golden rules for freezing

1 Put only good quality foods into the freezer. It can't work miracles; what comes out is only as good as what went in.
2 Handle food as little as possible and keep everything scrupulously clean. Freezing doesn't kill bacteria or germs.
3 Pack and seal food carefully. Food that is exposed to air or moisture will deteriorate and there is a risk of cross flavouring or transfer of smells from food that is inadequately wrapped.
4 Never put anything hot or even warm into a freezer. It will raise the temperature of other items in there and possibly cause deterioration. Cool food rapidly once it has been cooked or blanched, either in the refrigerator or in a basin of iced water.
5 Freeze food as quickly as possible so that it retains its texture.
6 Pack food for freezing in small quantities. It will thaw more quickly and you can get out more than one pack if necessary.
7 Follow the manufacturer's instructions on use of the fast-freeze switch and where in the cabinet to freeze down food.
8 Don't pack food that is to be frozen too closely together. Spread it out until it is frozen.
9 Move newly-frozen items from the fast-freeze area once they are frozen.
10 Remember to switch the fast-freeze control back to the normal setting once food has been frozen.
11 Don't keep opening the door of your freezer; it raises the temperature inside the cabinet. Decide what you want to get out and remove it quickly. Don't leave the door open while, for example, you remove 450 g (1 lb) of peas from a 2 kg ($4\frac{1}{2}$ lb) pack.
12 Use a freezer thermometer to help you maintain a steady storage temperature of −18°C (0°F). Move it around within the freezer to check that the temperature is maintained throughout the cabinet.
13 Label and date food so that you can rotate stock efficiently. Consider keeping a freezer log book to stop you going through the contents to see what you have stored.
14 Defrost the freezer when stocks are low and preferably on a

cold day. Do the job as fast as you can so that the contents can be put back quickly.

15 Tape the address of the service organisation to an inconspicuous spot on the freezer casing or inside the adjacent kitchen cupboard door.

16 Know what to do in emergencies such as a power cut or freezer breakdown.

17 If you are storing a large quantity of food, particularly expensive ingredients or dishes, consider taking out special freezer insurance to cover you against loss.

18 If your freezer is in a garage or outhouse, fit a freezer alarm to it so that you have immediate warning if anything goes wrong.

19 Put a plug with a built-in buzzer on the freezer to warn you if the fuse blows.

20 Your freezer should be full to keep running costs down. Fill gaps with basics like bread or, if you are running down your stock to defrost, fill the gaps with old towels or crumpled newspaper.

What not to freeze

Although most foods can be frozen in one form or another there are some items that just lose all their eating appeal if you freeze them. Here's a list of what doesn't work well.

Bananas turn black. This can be improved if the flesh is mashed with lemon juice and used in a recipe.

Caviar tends to go rather watery.

Cheese does not freeze satisfactorily if it is the cream or cottage type. Hard and blue-veined cheeses freeze well either in the piece or grated. Allow plenty of time for thawing. Curd cheese is best frozen in retail cartons sealed with tape. Camembert and other soft types should be matured before freezing.

Egg custard separates out although canned custard freezes satisfactorily.

Eggs cannot be frozen in the shell and when hardboiled they become tough and watery and the white turns grey. Egg whites and whole beaten eggs freeze well, but separated yolks become gummy unless salt or sugar is added.

Garlic-flavoured dishes tend to develop a musty flavour within a short time, which is a bit of a blow if you can't imagine your favourite terrine without it. Store for a short time only.

Gelatine Don't freeze dishes containing a very high proportion of gelatine, especially anything depending entirely on it as the stabilising agent, for example a moulded jelly. Gelatine-based recipes like soufflés and mousses freeze well for up to about one month.

Hollandaise sauce, mayonnaise and other egg-based sauces tend to separate.

Jam used in large amounts or as a filling has a tendency to liquefy and soak into the surrounding food.
Potato Some varieties of potato go leathery in pre-cooked frozen dishes unless they are mashed first.
Salad ingredients Lettuce, endive, watercress, spring onions, celery, cress, chicory, radish, cucumber and tomato go limp and mushy if frozen—and there's no way of getting their crispness back.
Sauces It's wise to keep the thickening of sauce to a minimum and use beurre manié when you reheat.
Single cream (that is, with less than 35 per cent butter fat) separates on thawing. It may go back to normal with excessive whipping but this is unlikely, so it's better not to freeze in the first place. See page 61 for directions for freezing double cream.
Yoghurt that is homemade or natural separates when frozen. Flavoured sweet yoghurts show little or no deterioration.

FREEZER FACTS

What type will I buy?

The type of freezer you choose will depend on the amount of space you have to put it in and what kind of food you plan to store in it. Both chest and upright freezers have their fervent adherents; the important thing is to buy the type that suits you best.

Upright freezers

The same shape as a full-size refrigerator, these range in size from small, table-top models to large upright ones. They have front opening doors—large freezers sometimes have two doors—take up less floor space than chest freezers and are less obtrusive.

Upright freezers, fitted with their shelves and/or pull-out baskets, are easy to load and unload. In some models each shelf or basket is fronted by its own flap which shuts out warm air when the door is opened. Upright freezers need defrosting more often than the chest type—although frost-free models are available, they are more expensive both to buy and to run.

Chest freezers

Size for size, chest freezers are cheaper than upright models, and running costs are lower. They have an additional advantage too: more food will fit into the same amount of space, since large items can be accommodated with small ones tucked in round them, unrestricted by shelves and baskets. However, they do take up considerably more floor space; are more difficult to keep organised, and small people and those with bad backs will find it difficult to reach down to the bottom of the cabinet.

Refrigerator/freezers

A unit which incorporates a fridge and freezer, one built above the other, is popular in small kitchens with little space. The refrigeration and freezing capacities are roughly equal in many models, although it is also possible to buy units which have bigger freezers than refrigerators, and vice versa.

What to look for when choosing

Before buying, check the models you are interested in carefully. See how easy it will be to open the door when the freezer is in

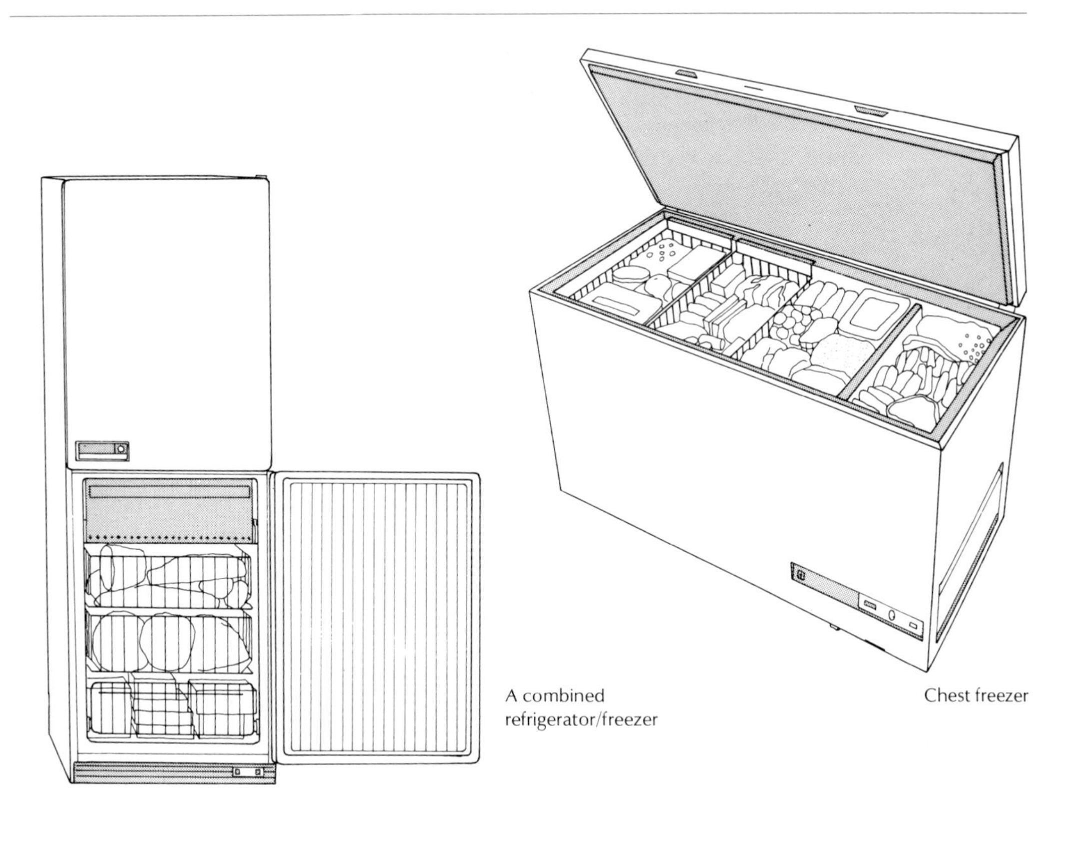

A combined refrigerator/freezer

Chest freezer

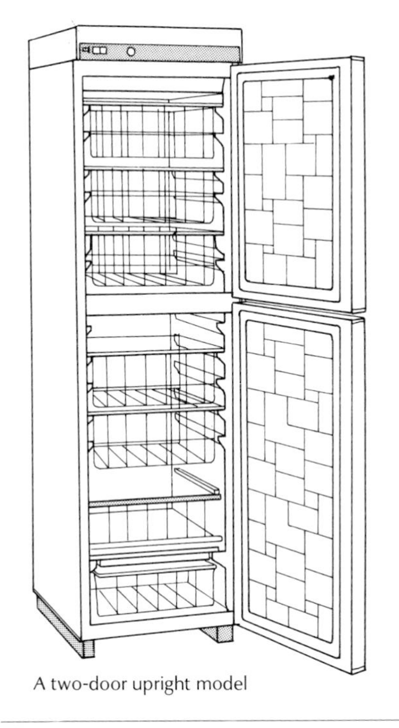

A two-door upright model

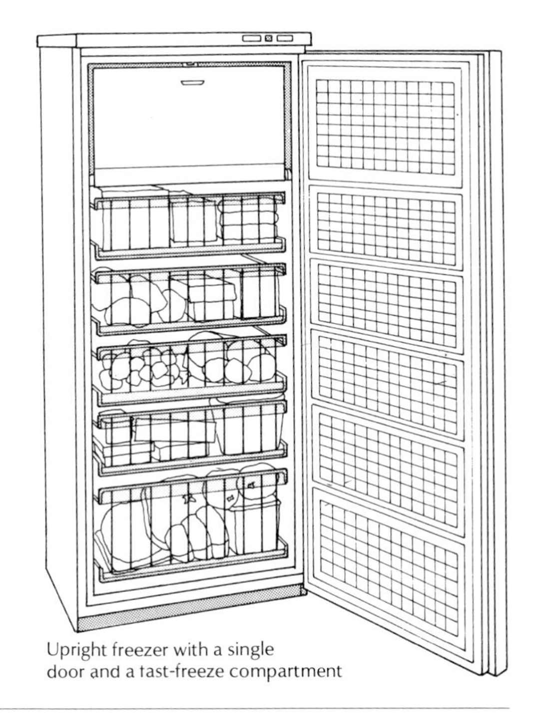

Upright freezer with a single door and a fast-freeze compartment

position in your home—it is sometimes possible to buy the same model with the door hung on the other side.

Can you reach the top shelf of an upright model and the bottom corners of a chest easily? Look at the position of the controls: are they simple to use and to understand? Could they be altered by an inquisitive child?

Has the freezer got a lock—probably necessary if you are going to keep it in a garage or shed—and do you need an interior light?

A drainage hole is useful in a chest freezer—check it has one so that you don't have to bale out when defrosting.

Look too for rollers, which enable you to slide the freezer out to clean behind and beneath it, or for adjustable levelling feet if your floor is uneven. Check what is the maximum amount of fresh food that can be frozen in 24 hours.

The star symbols

All freezers must carry this star symbol ✱✱✱✱ which indicates that they are capable of freezing down fresh food. An appliance with three stars or less is suitable only for storing ready-frozen food:

* for 1 week
** for 1 month
*** for 3 months

The fast-freeze switch

The fast-freeze switch works by overriding the thermostat. It allows the temperature in the freezer to fall well below −18°C (0°F)—the normal storage temperature (which is controlled by the thermostat). The food freezes faster in the colder temperature and because smaller ice crystals are formed the texture of the food will be better when thawed. Reducing the temperature means, too, that the food already in the freezer gets colder and will be less affected by the higher temperature of the fresh food. Turn on the fast-freeze switch about six hours before you plan to put in food to be frozen, then leave it on for up to 12 to 24 hours to freeze the food (see page 29).

It is possible to freeze down food, especially in small quantities, without using the fast-freeze switch, but the faster the food is frozen the better it will taste when thawed.

What size freezer?

The freezer capacity is quoted in either cubic litres or feet. (To make a rough comparison divide the number of litres by 30 to give the number of cubic feet (1 cubic foot = 28.3 litres).) Some manufacturers quote capacity in gross volume and some in net, so watch out for this.

How large a freezer you need depends on a number of factors: for instance, how often you shop; whether you plan to cook for your freezer or merely to store basic ingredients; how many people there are in the household, and whether you will be freezing your own produce.

It is generally suggested that you should allow 56 litres (2 cubic feet) per person plus 56 extra litres (2 cubic feet) but do think carefully about your particular needs before deciding. Talk to friends who own freezers to see whether they are satisfied with the size they have. If in doubt buy a model that is 28–56 litres (1 or 2 cubic feet) larger than you think you need—you will probably use the extra capacity.

Where to put a freezer

The most convenient place to site a freezer is obviously in, or near, the kitchen. But even if it appears that you can fit a freezer in an ideal site, you might discover a few snags. Chest freezers take up a lot of floor space but are unlikely to provide an equivalent

For reasons of space, many people prefer to keep their freezers outside, in the garage, for example. If this is done, the freezer should be protected from slight rising damp by raising it on wooden blocks to improve air circulation. A coating of silicone wax polish will also shield the body of the freezer against rust.

amount of worktop area—as soon as you start to prepare food on top of the freezer you're bound to need something out of it.

Upright freezers may need more than their width for you to be able to open the door properly: on some models the door must be opened more than 90° before you can pull out the baskets. Bear in mind also that you need to allow about a 5-cm (2-inch) gap at the back for ventilation. The overall design of your room may be spoilt, for example, by tall uprights and fridge-freezers placed in the middle of a wall—they are best sited at the end of a run of units. Check that the height will not take light from the kitchen.

Of course a freezer can be sited anywhere in a home or even in a garage or shed, but bear in mind that it does make a certain amount of noise when the motor is on which may rule out a dining room or bedroom. And you'll want the freezer to be reasonably accessible so that you don't have to walk upstairs or out to the garage every time you want a couple of beefburgers.

If you are installing a freezer upstairs be careful that it is not tilted to an angle of more than 30° from the upright or you may get an airlock in the coolant (see also page 21, moving house). A chest freezer, which will be very heavy when full, should always be positioned across the joists if upstairs, otherwise it could damage the floorboards.

It is best to put a freezer sited in a garage or shed on blocks (strong bricks or solid pieces of wood) to prevent it getting damp and rusting and it should be checked often for such damage. It will probably be cheaper to run, though, than if it was inside a warm house.

Any freezer which is kept in an out-of-the-way place should be fitted with a freezer alarm so that you get immediate warning if anything goes wrong.

Running costs

Servicing and repair bills are of course hard to estimate, but generally freezers are reliable and in fact may last for years without attention (see page 18, maintenance contracts).

It is possible to get a rough idea of the running costs of a freezer. On the rating plate (usually on the back of the appliance) you'll find what the wattage of your model is. (Look at the figure before the 'w' or 'kW'. If it is in watts divide it by 1000 to give the kW rating.) Multiply this figure by the current cost of a unit of electricity—this will give you the running cost for an hour. However, the freezer is only using electricity when the compressor is pumping the coolant round the system, so the actual running cost will be much lower than this. You can keep the costs down by taking a few simple measures:

1 Don't site the freezer near a central heating boiler, hob or oven which gives off heat, or where direct sunlight will fall.
2 Open the freezer door/lid as little as possible.
3 Only put cold food into the freezer.

Secondhand freezers

People sell freezers for a number of reasons: they may be going abroad; moving to a smaller house; need a larger freezer, or buying a fridge/freezer. A secondhand freezer can be a good buy but you should check it properly before handing over your money. See if the appliance is a freezer with a star rating and not merely a frozen food conservator. Find out how old it is; if over five years or so you could be reaching the point where things *might* start to go wrong. Check with the manufacturer or your local electricity board that spare parts and servicing are available for the particular model. Look for signs of rust. Make sure the controls work (beware of freezers that aren't in use or haven't been for some time) and test whether the door seal is tight by shutting a strip of paper in the door and seeing if it moves around easily. If you are satisfied buy the freezer, but remember to take all the right precautions when moving it from its old to its new home (see page 21).

Some shops sell reconditioned freezers which are often very good value and usually come with a three- or six-month guarantee. This is possibly the best way to buy a secondhand freezer since the rules of the Sale of Goods Act (1979) apply—it must be of merchantable quality and fit for purpose.

Installing a freezer

All you need to run a freezer is a 13-amp socket outlet. But there are a number of things you must do before you plug in your freezer for the first time.

First of all make sure there is a socket outlet close—you don't want metres of extension lead running through your home. It is worth paying to have a suitable socket fitted and at the same time putting the freezer on a separate circuit from the rest of the house. This means that you can switch off all electricity except the freezer supply when you go away for a few days.

If you are plugging a freezer into one half of a double socket outlet do not use the other half for appliances, such as food processors or coffee grinders, which are plugged in and unplugged frequently, because you may accidentally unplug the freezer. It is a good idea to cover the socket switch and the freezer's plug with tape to deter unplugging. Alternatively, fit a plug which incorporates an automatic buzzer if the socket is switched off.

Make sure your freezer is in position with the correct air gap behind and around it (check the handbook) and that the feet are either levelled up if they are adjustable or packed up with small pieces of hardboard. Use a spirit level.

Wipe the interior of the freezer with a solution of 15 ml (1 level tbsp) bicarbonate of soda to 1 litre ($1\frac{3}{4}$ pints) warm water and dry it thoroughly. After the freezer has been moved into position, allow the oil in the compressor to settle for several hours before switching the unit on (leave the door open).

Once you have turned it on, put in some food and it is running normally, check that the freezer is maintaining the right temperature. Use a freezer thermometer and move it to a different position every 24 hours to see if there are any spots which are particularly

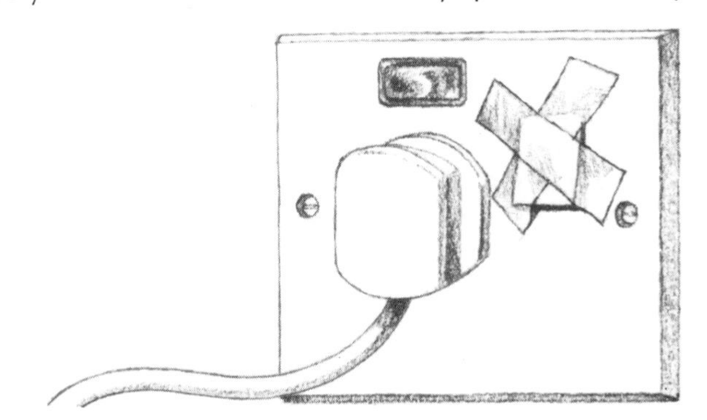

If the switch of your freezer is accidentally turned off it can prove very expensive in terms of lost contents. If your children are inquisitive toddlers, make sure the switch is secured with adhesive tape. It will stop you accidentally switching off the wrong appliance too!

cold or warm. Set the thermostat so that all parts of the freezer maintain a temperature of −18°C (0°F) or less. There is no need to keep the freezer colder than this except when you are using the fast-freeze switch (see page 29)—lower temperatures cost more to run.

Defrosting

Defrost your freezer when the frost on the shelves reaches a thickness of 0.5 cm ($\frac{1}{4}$ inch). Follow the manufacturer's instructions for the procedure. You will need a plastic spatula to tap the frost loose as it melts, some old towels to mop up with and bowls of hot water to speed up the process. You can use a hand-held hair drier to make defrosting even quicker but take care not to let it touch any frost or water and keep it moving rather than blasting continuously at one spot.

Try to defrost when your stocks are low or when it's a cold day. Put the contents of the freezer in the refrigerator, a cool box or wrap it in newspapers or old blankets.

When the freezer is free of frost wash it out with a solution of 15 ml (1 level tbsp) bicarbonate of soda to 1 litre ($1\frac{3}{4}$ pints) warm water and dry it thoroughly before switching on again.

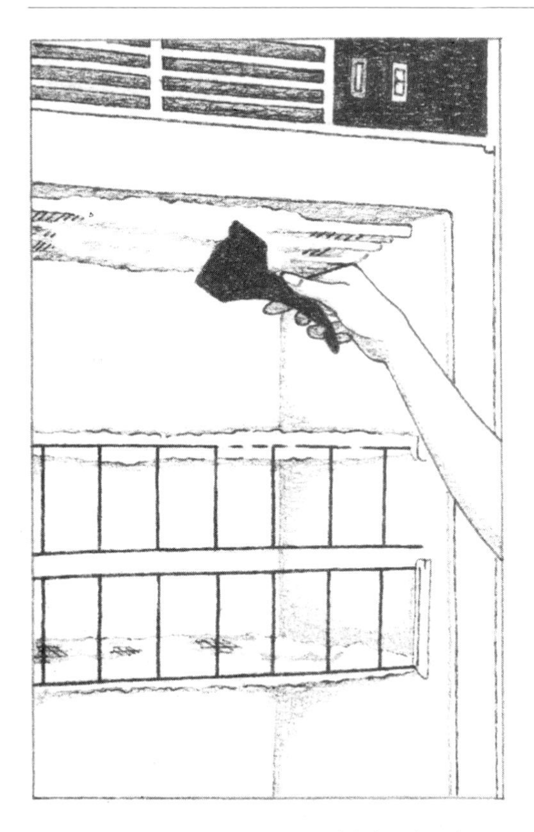

Use a plastic or wooden spatula to speed defrosting. Never scrape out ice with a metal implement as this will damage the inner surface of your freezer.

Avoid a flood by using a sheet of foil to funnel melting ice into a bowl.

Maintain the outside of the freezer cabinet by wiping it over occasionally with a solution of washing-up liquid and warm water. Always dry it thoroughly. Use the crevice tool on the vacuum cleaner to remove dust from the network of pipes at the back, taking care not to damage them. If the freezer is in a garage or shed polish it from time to time with a silicon all-purpose polish to help it resist damp.

FREEZER EMERGENCIES

Even if you're obeying all the rules about freezer care and freezing, things can go wrong. A blown fuse while you're on holiday, a power cut or compressor failure can mean that all your carefully prepared frozen dishes and produce are at risk of going to waste. But you can take some precautions to minimise losses.

Precautions

Insurance

The contents of your freezer could be worth considerably more than various odd pieces of furniture which are covered by your household contents policy, so it makes sense to insure them. If you cook a lot for your freezer or buy meat or fish in bulk the value of the contents may be several hundred pounds—a sum well worth paying a premium for.

A number of insurance companies offer freezer insurance as part of a household contents policy while others issue a separate policy. In either case, and especially the former, do read the fine print carefully and find out exactly what is covered. It may be that the value of the freezer contents is fixed too low or the policy may incorporate exclusion clauses stating, for example, that if a power cut is caused by industrial action they will refuse to pay out.

Also check exactly what is covered in terms of freezer repair, theft (important if the freezer is in a shed or garage) and accidental switching off of the power supply (which is usually not covered). Some companies offer the loan of a replacement freezer if yours is out of action.

Find out whether the premium will increase as your freezer gets older and if the insurance has to be taken out for a fixed number of years. If you have difficulty finding a company offering the kind of freezer insurance you want, contact Ernest Linsdell Ltd, Commonwealth House, 1/19 New Oxford Street, London WC1A 1NB.

Guarantees and maintenance contracts

When you buy your freezer it will have a guarantee, usually of one year. Some manufacturers allow you to extend the guarantee to five years on payment of an extra sum at the time of purchase. Before you pay this, check that the extended guarantee covers both parts and labour. Sometimes only the compressor is covered, with the other parts of the freezer guaranteed for only one year.

When the guarantee period expires you will have to decide whether or not to take up the manufacturer's offer of a maintenance contract. Look carefully at the terms of the contract to see what is covered—both parts and labour should be if it is to be worthwhile. Make sure too that a daily emergency service operates. You may decide it is better value to save the annual cost

A fridge/freezer is a popular combination as it takes up minimum floor space and provides good storage capacity.

of the maintenance contract and put the money towards a repair when you need one, or a new freezer.

Power cuts

All freezer owners dread power cuts. You can make your way around the house by torchlight or even using candles (which can be quite romantic!) and you can cook on a camping stove. But the freezer has no alternatives to fall back on.

Most power cuts do not last for long, however, and provided you follow these rules the food in your freezer should be safe.

1 If there is advance warning of a power cut at least 6 hours before it occurs, turn on the fast-freeze switch. Before you do this, make sure the freezer is as full as possible. Fill any gaps with rolled-up newspaper, old towels or plastic boxes filled with water.

2 Do not open the door during a power cut as this will let warm air into the freezer.

3 Cover the freezer with a blanket or rug to increase its insulation but make sure you leave the condenser and pipes on the back uncovered.

4 After the power has been restored leave the freezer on fast-freeze (or switch to it if there has been no advance warning) for at least 2 hours.

Provided you take these precautions food in a chest freezer will be undamaged for about 48 hours if you have had advance warning and 30–35 hours without. An upright freezer will keep its contents safe for about 36 hours with advance warning, 30 hours without.

Food in a refrigerator which has been without power for over 24 hours will not last as long as the recommended storage time. If the power cut is a long one see if friends or your local butcher can store your frozen food until power is restored, although they themselves may be affected of course.

Breakdowns

Freezers are, generally, reliable but they may break down from time to time. Before you call a service engineer check the following points:

1 Is there a power cut? Switch on a light.

2 Is the socket outlet working? Plug in another appliance and check.

3 Has the fuse in the plug blown? Change it and see.

4 Is the compressor working? Switch to fast-freeze and listen.

If the freezer seems to be too cold you may have left the fast-freeze switch on or accidentally knocked the thermostat setting.

If it sounds very noisy check that the freezer has not been pushed too close to the wall and that it is level.

You may be able to repair a faulty door seal or perished rubber pads by the compressor yourself—you should be able to buy the spare parts from the manufacturer. If you do need a service engineer and he or she isn't able to come that day, transfer food to friends' freezers if possible.

Reducing losses

It is not always necessary to throw out all the food from your freezer if there has been a lengthy power cut or breakdown. Some foods can be refrozen and others can be cooked, then frozen. As a rough guide, if the temperature of the freezer has not gone over 0°C (32°F) the food can be refrozen by switching on the fast-freeze control for a few hours. The items which are most likely to be affected by thawing are meat, poultry and fish so if these seem to have thawed out too much, cook and then freeze them. Be sure to thaw any you do not cook in a refrigerator and cook as soon as they are thawed.

Moving house

Freezers don't take kindly to being moved so it's important to check that a professional removal firm is prepared for the responsibility of moving your model. If you are moving the freezer yourself do take particular care with it.

It is best to move a freezer when it is empty, although some removal firms have the facility to plug a loaded unit into an electrical supply in their van. If you want this done the freezer must be the last item loaded on to the van and the first off into your new home. Bear in mind that a freezer full of food is very heavy and might be damaged because of this. It is really better to use its contents to see you through the frenzied days before the removal and transport it empty.

When moving a freezer it should be kept as upright as possible and never tipped to an angle of more than 30° (this could cause an airlock in the cooling system). When it is unloaded and installed in your new home switch the freezer on to check that all the lights are working then turn it off and let the oil in the compressor settle down (see page 16) before running it normally. When you sign the removal firm's document stating that the job has been completed, write a note on it to the effect that the freezer has not yet been tested so that you can take action if you find that the pipes have been broken in transit.

ALL WRAPPED UP

Good packaging is vital if food in the freezer is to remain in optimum condition. Freezing converts the water content of the food to ice crystals which must be retained: they are converted back to moisture when the food is thawed. Badly-packed food will dry out causing the white patches known as 'freezer burn'. There is also the risk that strong-smelling foods will transfer their odours to other foods. Although it is important to store all frozen food correctly, any which is to be stored for a long time must be extremely well wrapped since it will tend to get pushed around when other items are put into and removed from the freezer.

What you need for packing

Aluminium foil

Aluminium foil is very useful for wrapping because it can be shaped to cover awkward items and pressed close against the food to ensure air does not get in. Use a single layer of the standard thickness or a double layer if you are using thin kitchen foil. Seal it by folding the edges closely over each other, if neces-

Pliable wraps are the most versatile for any freezer owner to use.

sary using freezer tape to hold it in place. Aluminium foil should not be used for wrapping acidic fruits which may react with it. If you think the foil is likely to become punctured in the freezer, overwrap large items with a polythene bag (this need not be freezer gauge as the foil will protect the food).

Aluminium foil is also useful for lining casserole dishes when preparing food. Once the dish has been cooked and frozen the foil lining containing the food can be removed from the casserole. Overwrap the parcel and store in the freezer until needed when the food can be unwrapped and replaced in the casserole (without the foil liner) for reheating and serving.

Standard aluminium foil is tough enough to be re-used for short-term storage, but it must be soaked clean and dried thoroughly first.

Polythene bags

Both polythene bags and sheeting can be used for freezer wrapping; the bags, available in a variety of sizes, are easier to use than sheeting which needs careful wrapping and sealing. Use heavy-gauge polythene which will protect the food, rather than the thin bags used for packing sandwiches and keeping food in the fridge. Most foods can be stored in polythene bags, but liquids and solids-plus-liquids such as stews, should be frozen in a polythene bag placed in a container, such as a cereal packet, icing sugar box or plastic container. The square or rectangular shapes produced are compact to store in the freezer.

Air should be removed from the bags before they are put in the freezer. Either suck it out with a straw or vacuum pump or put the

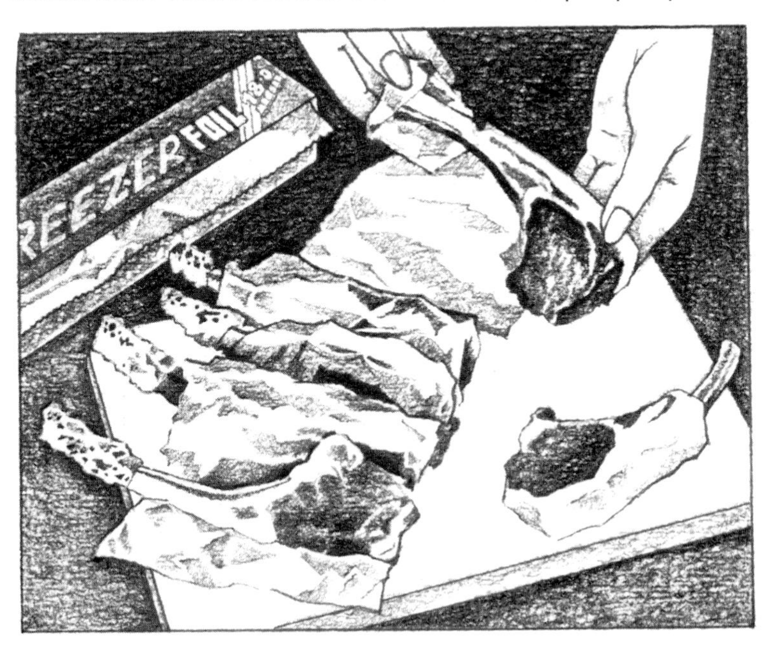

When packing cutlets pad any protruding bones with foil to prevent them puncturing the foil pack. Pack cutlets and steaks concertina style to keep each one separate.

bag into a basin of cold water which will force the air out. If you use this method be careful not to get any water inside the bag, and dry the outside thoroughly to prevent frost build-up in the freezer. Seal the bags with freezer tape or twist ties.

Cling film/freezer wrap

Plastic film tends to cling to itself and also to plastic and metal so is very useful when wrapping food for the freezer. Use it as a lining if you want to pack acidic fruits in foil and also for wrapping individual portions of foods which can then be stored together in a polythene bag and removed one at a time. (It can be doubled to cover casseroles and ramekins—they can be put straight into a microwave oven if a hole is first pierced in the film.)

Aluminium foil dishes

Foil dishes come in a range of sizes and shapes and you can cook, freeze and reheat food in them. Used and cleaned carefully they can be recycled several times, but note that foil dishes cannot be used in a microwave oven, and lids should not be re-used.

The longer frozen food is stored, the more important good packaging becomes. Keep a ready supply of moisture and vapour-proof containers suitable for the freezer.

Plastic containers

Plastic containers are more expensive than other types of freezer packaging but will last for years. Some varieties may lose their air-tight seal after a while and need sealing with freezer tape to prevent air affecting the contents—good quality brands such as Tupperware carry a useful 10-year guarantee. While you may want special plastic shapes for some foods, ice cream bombes for instance, square and rectangular containers make better use of freezer space than round ones. If you eat commercial ice cream, 'free' containers will multiply in your home and these can be used very successfully for food storage.

Other packaging

Packaging containers from some bought foods are useful in the freezer. Good examples are margarine tubs, yoghurt and cream pots and foil dishes. Make sure they are scrupulously clean and do not expect them to survive more than two or three sessions in the freezer. Use freezer tape to ensure an airtight seal.

Specially toughened glass dishes are handy for mousses and desserts which are to be served in the dish, but they must be freezer proof.

Sundries

In addition to wraps, bags and boxes you will also need sheets of waxed or non-stick paper; plastic- or paper-covered wire twist ties which can be bought ready-cut or on a roll; freezer tape; labels; a chinagraph or waterproof felt pen; a vacuum pump, or supply of straws, and possibly an electric heat sealer.

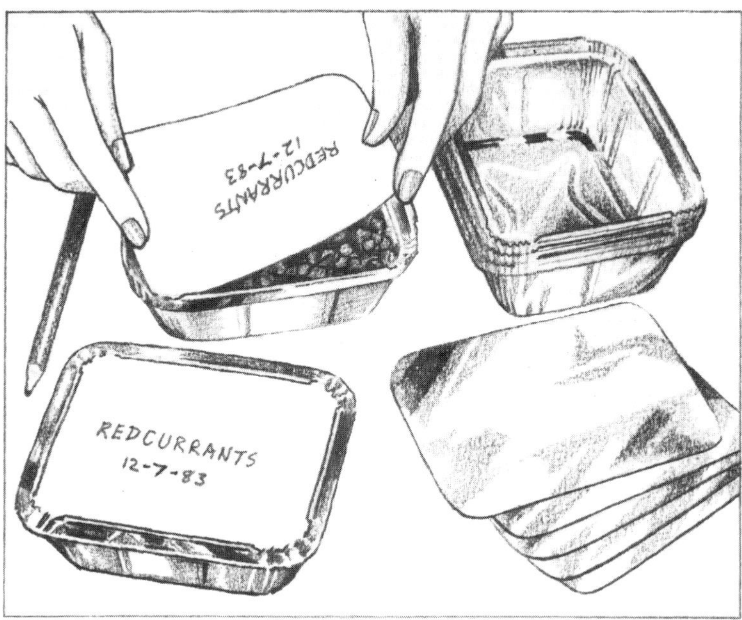

When using foil dishes remember that the lid goes *foil down*.

Packing in manageable quantities

It is very important to pack food so that you can easily remove the amount you want from the freezer. Chops and pancakes, for example, should be frozen with paper interleaved between them, then wrapped, so that you can remove only the number you require. Casseroles and pâtés are best frozen in relatively small quantities—you can then thaw as many packets as you need

When wrapping solid pieces of fish or meat use the butcher's method, shown here, and seal the ends with freezer tape.

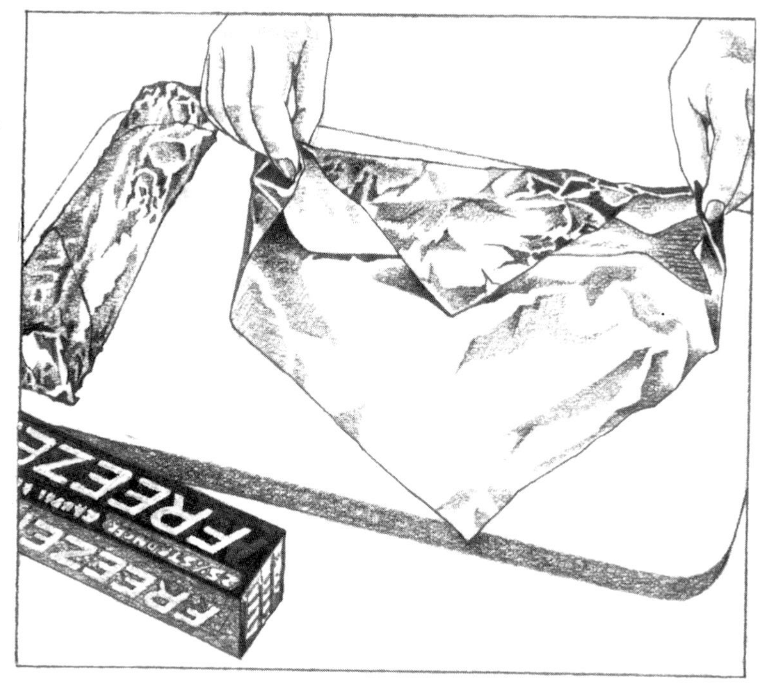

It is best to pack cake, meat slices and sandwiches etc. in portions for easy serving.

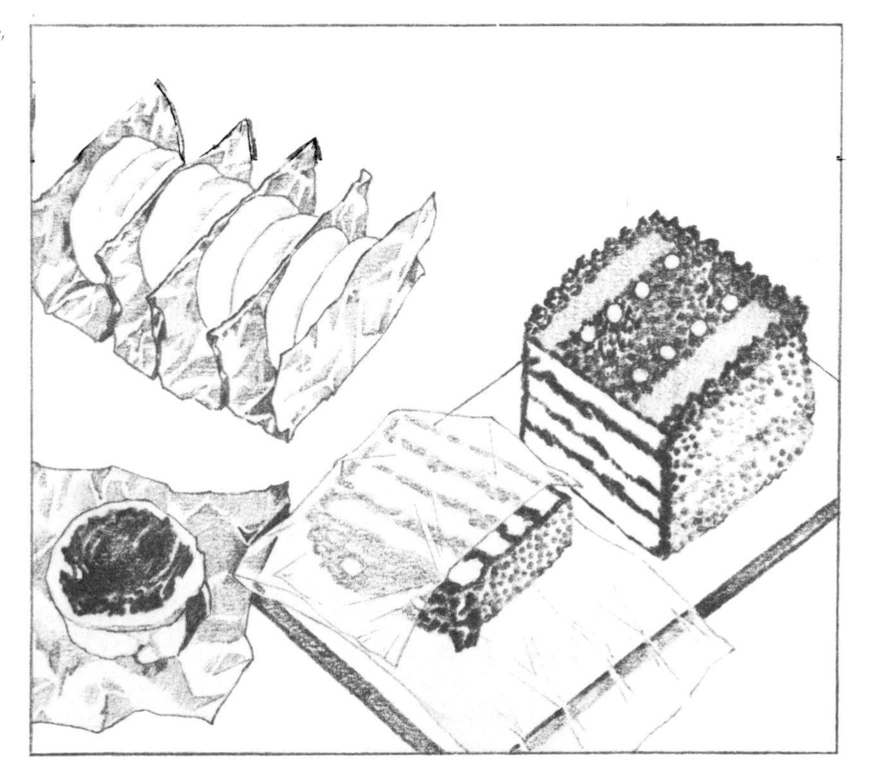

rather than try to saw a giant lump of frozen stew or bolognaise sauce in half. Soups, stock and sauces should not be frozen in quantities larger than 300 ml ($\frac{1}{2}$ pint). Freezing liquid in ice-cube trays and storing the cubes in polythene bags is a good way of retrieving a minute quantity when required. With free-flow items such as peas or beans you can just pour out the amount needed.

Sealing

Provided you have first extracted all the air (see pages 23 to 24), twist ties are an excellent way of sealing bags. Freezer tape is another good sealing method. It has a special adhesive that sticks at low temperatures, unlike ordinary sticky tape, and is available in white, which you can write on and save the extra cost of labels.

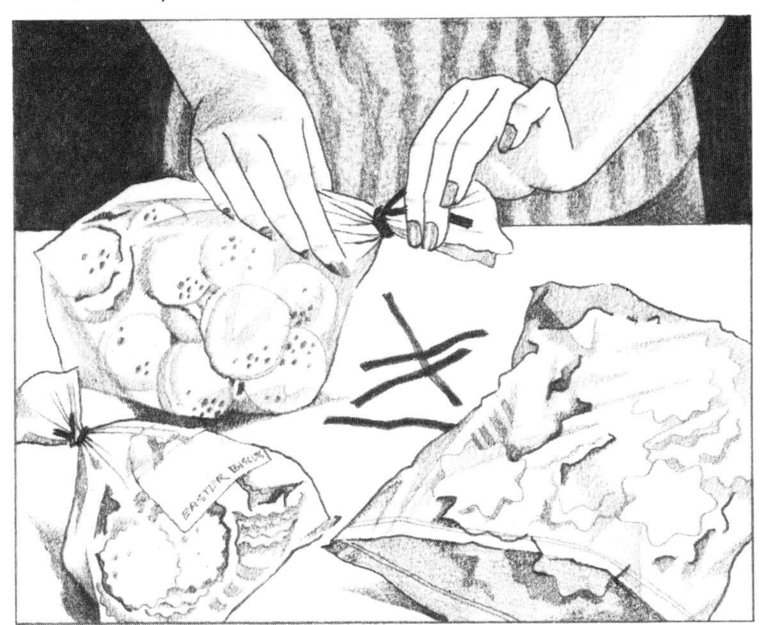

Seal polythene bags with twist ties or use special freezer tape which does not lose its adhesive qualities at sub-zero temperatures.

Heat sealing is a very secure method of sealing polythene bags and is particularly useful if you want to use the packet as a 'boil-in-the-bag' item. You can buy special electric sealers for freezer bags, or run the tip of your electric iron along the bag opening, over which you have placed a sheet of brown paper. It is probably not worth investing in a special sealer unless you are sure you will use it regularly.

Labelling

It is essential to label everything that goes into your freezer, unless it is a commercial packet which clearly states what the contents are. However convinced you are that you will be able to tell bortsch from blackcurrant sorbet when you put them in, after a few months they will seem identical. You will need to use a china-graph pencil or waterproof felt-tipped pen for labelling, so that the information does not become rubbed or wiped off. Some polythene bags have a white label printed on them; on other items you will need to use a special freezer label with an adhesive which

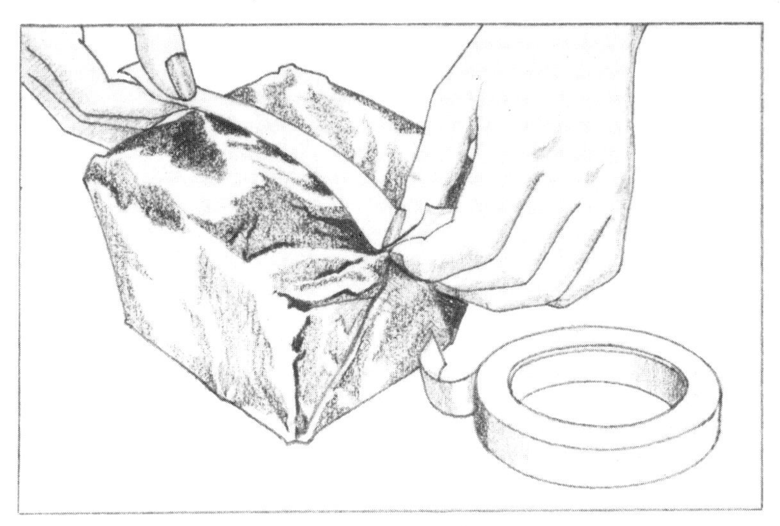

A good supply of freezer tape is essential for sealing all types of packages.

When labelling make sure your marker is waterproof. Ideally, a label should give the date of freezing, the contents and the number of portions.

sticks at low temperatures. On each packet write the contents, quantity and date of freezing. If other people are going to cook and eat the food it may be worth including brief instructions about thawing, reheating and any extra ingredients which should be added. You can buy freezer labels in different colours so that you can code things for easier finding—for example red for meat, green for vegetables, and so on.

HOW TO FREEZE

Freezing is a very simple process, based on how quickly heat can be taken out of food (*not* cold pumped in), but unless you grasp the basics, it's possible to make some equally simple mistakes. These rarely make the food inedible, but they may affect its taste, texture and—most vital of all if you're feeding a young growing family—its food value.

Most foods are largely made up of water, and even something as seemingly solid as lean meat contains about 70 per cent of it. All freezing does is convert this water to ice crystals. Quick freezing results in tiny ice crystals, retained within the cell structure, so that on thawing, the structure is undamaged and the food value unchanged. But slow freezing results in the formation of large ice crystals, which damage the cell structure and cause loss of nutrients. As this damage is irreversible (slow-frozen food shows loss of texture, colour and flavour when thawed), you'll see why it's vital to follow the manufacturer's freezing instructions implicitly. Foods like cucumber and strawberries,. which have a high water content, never freeze successfully as harvested because the tiniest crystal formation breaks down their delicate structure.

First essential is never to freeze more than one-tenth of your freezer's capacity in any 24 hours or else not only will heat be absorbed by the freezer's refrigeration system, but also by food already frozen. If, for instance, you have a 50 kg (110 lb) freezer (approximately 170 litres or 6 cubic feet), you should only ever freeze about 5 kg (11 lb) of food at a time. It's *possible* to freeze more, of course, but because the addition of the unfrozen food pushes up freezer-temperature, the results are going to be *slow-frozen*—the very thing you want to avoid.

How should I use the fast-freeze switch?

Not all manufacturers give clear instructions on how to use the fast-freeze switch, and if this is the case with you take the following as a guide.

For small amounts; for example, one loaf plus one small casserole
It is probably not necessary to touch the switch at all, simply pop the food in. This should be safe for up to, say, four items but not including anything as dense as a leg of lamb.

For a fairly small amount; for example, four casseroles and three pies
Put the fast-freeze switch 'on' for about 2 hours before you put the

food in. Leave it on for about 4 hours more, until the food is really solid.

For a large amount; for example, a half-carcass of fresh meat
Put the fast-freeze switch 'on' for about 6 hours beforehand to ensure the freezer is really cold. Put the packaged meat in and leave the switch on for a further 12–24 hours, depending on the load.

Can I put unfrozen food straight into the cabinet?
Some freezers have a separate freezing-in compartment. The temperature in this compartment will not be any different from that in the main part, but the division will prevent warmish food coming into contact with what is already frozen and starting a slight thaw. If your freezer has no separate compartment, put the food against the sides or base, or on a shelf that carries the evaporator coils—that is, in the coldest possible part of the freezer.

Solids

Package solids tightly, so that you expel as much air as possible. This is easy if you are wrapping something in aluminium foil or freezer wrap, which fits where it touches, but it's not so simple if you're filling a rigid container, and run out of food halfway up. You can't squeeze the surplus air out, but you can fill up the vacant space with some crumpled foil or non-stick paper. Half-empty containers waste freezer space, though, so try to avoid this.

It's even more difficult to expel air from a polythene bag, especially when the contents are awkwardly shaped or easily

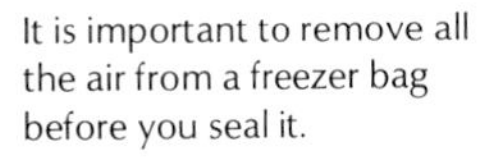
It is important to remove all the air from a freezer bag before you seal it.

When freezing liquids, or solids in liquids such as casseroles, pack them allowing a headspace for expansion.

broken. It's quite simple if you take one of the children's drinking straws and suck the surplus air out. Alternatively, you should lower the full bag into a bowl of water, so that the water-pressure can force the air up and out (remember to dry the bag before freezing, so you don't find yourself having to chip it out when you need it).

Liquids

These are a very different proposition: instead of squeezing air out, it's essential to leave at least 1 cm ($\frac{1}{2}$ inch) of 'headspace' in—up to 2.5 cm (1 inch) for about 600 ml (1 pint). This is because water expands one-tenth on freezing.

Unless you leave room for expansion, items such as soups, sauces and fruits packed in syrup will push off their lids; worse still, if you've used a screw-top jar (difficult to remove food from, and unless the glass is specially toughened, frighteningly fragile), food could shatter its way out. Bear this in mind, too, if you're attempting to chill a bottle of wine or beer in an emergency, because if you leave it too long, it will be splattered around the freezer by the time you get to it. Never do this with any fizzy drinks.

If you do use a glass container, make sure it has straight sides, which make it easier to get the contents out again, and leave 2.5–5 cm (1–2 inches) headspace.

Solids-plus-liquids

Combinations of solids and liquids, stews and casseroles for example, or fruit in syrup, should if possible have a layer of liquid on the top, with no pieces of food sticking out. Remember to leave 1 cm ($\frac{1}{2}$ inch) headspace. Solids which rise above the surface of the liquid, such as fruit salad, need an inner covering of crumpled non-stick paper or foil before completing the wrapping.

If a container is only part-filled, use crumpled greaseproof or waxed paper to fill the air spaces.

Preforming

When using polythene bags for liquids, use the right size bag and fit it into a regular-shaped container with straight sides such as an empty sugar or cereal carton or, of course, a plastic box will do the job efficiently. Freeze until solid. Ease the bag out of the pre-former, then seal, label and overwrap if necessary. Foil is also ideal

Not everything needs a special container. Liquid foods such as fruit pureés, casseroles and stocks etc., can be frozen in polythene or foil inside a preformer.

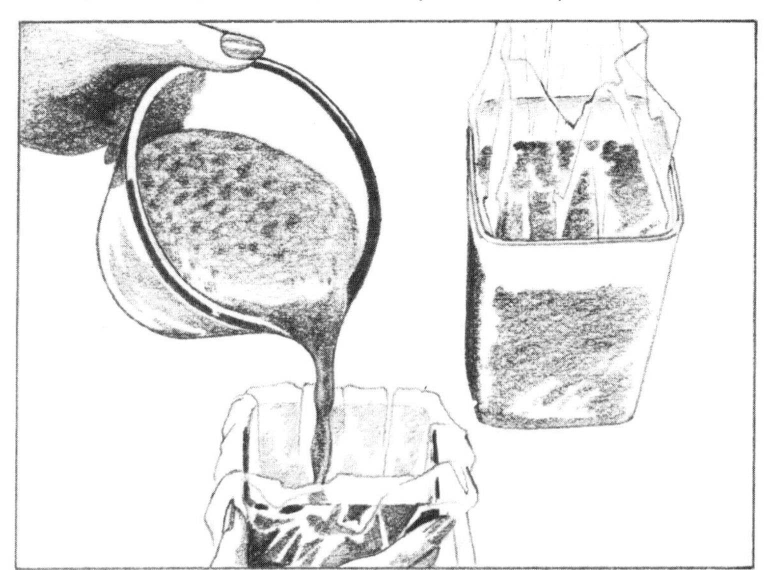

Three scrumptious pastry treats from the freezer—satisfying Sausage and Bacon Jalousie, Bacon Quiche and Sweet Orange Aigrettes with Apricot Sauce.

Sifter

for this method if you want to end up with a neatly-shaped casserole instead of a space-wasting lumpy bundle. Of course, you can freeze food in its casserole dish but if you want to keep the dish in circulation, preforming is the answer. Line the casserole with foil first, leaving a good margin for wrapping over; spoon in the contents. Once they're frozen solid, slip out the package and

A quick dip in warm water will loosen the frozen contents of this bag enough to release it from the preformer.

fold the foil over, then overwrap with a polythene bag before freezing. Slip the unwrapped package back into the original casserole dish for reheating (if the foil tends to stick, dip the package in warm water for just a minute or two). (An alternative method is to cook the food in the casserole without foil, freeze it, then dip the casserole in warm water just long enough to loosen the contents, which can then be removed and wrapped in the usual way.)

Only proviso: unless the casserole dish is straight sided and lipless, the contents may not 'slip out' as obligingly as we're suggesting.

How to freeze vegetables

Remember that food can only come out of the freezer as good as you put it in. Freezing doesn't *improve* quality, so don't waste valuable freezer-space on anything else than the best. With vegetables this means freezing them just as they reach maturity; the carbohydrate content of peas and beans, for instance, changes from starch to sugar at this point.

Blanching is essential when dealing with vegetables: its purpose is to destroy the enzymes present, preserve the colour, flavour and texture and, to a certain extent, to reduce the micro-organisms

Wholesome Vegetable and Bacon Broth is a useful freezer item, and so too are exotic Mushrooms à la Grecque and elegant Duchesse Potatoes.

Most vegetables need blanching before you put them in the freezer. On the right of this illustration some runner beans are about to be plunged into boiling water in a blanching basket. On the left, the drained beans are being plunged into ice-cold water to prevent over-cooking.

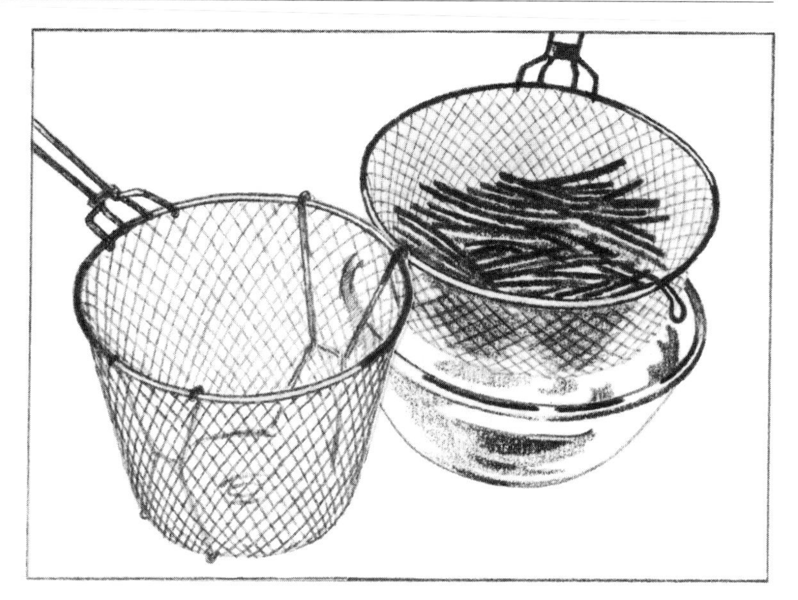

present; it also helps to retain vitamin C. The blanching water can be used 6 or 7 times, thus achieving less loss of minerals and a vitamin C build-up in the blanching water. After this time, replace with fresh water. The vegetables should be prepared as you would for normal cooking, and then blanched in not less than 4 litres (7 pints) boiling water to each 450 g (1 lb), with 10 ml (2 level tsp) salt. The water must return to the boil in 1 minute after the vegetables are added. Don't attempt to blanch more than 450 g (1 lb) at a time. Follow the times in the chart on pages 35 to 37. Calculate from the time the water re-boils. A blanching basket makes life easier.

After blanching, remove vegetables at once and plunge them into *ice-cold* water (add ice cubes to the water) to prevent over-cooking and to cool as quickly as possible. Cooling time is usually the same as the scalding time. For each batch use a fresh lot of iced water. Drain in a colander immediately the vegetables are cool.

Careful timing ensures the best results, so a 'pinger' or a watch with a minute hand is really the only answer.

For the sake of convenience in the chart we have included avocados, tomatoes and chestnuts under the heading of vegetables, since they are mostly used as savoury items.

Note: In an emergency—when for example you are dealing with a glut crop—vegetables can be frozen unblanched but this method is not to be generally recommended. A notable change in the eating quality of unblanched Brussels sprouts can be detected after only 3 days; in broad beans after 3 weeks; in runner beans after 1 month; in peas after 6 months.

Vegetable	*Preparation*	*Blanching time*
Artichokes, globe	Remove all outer coarse leaves and stalks, and trim tops and stems. Wash well in cold water, add a little lemon juice to the blanching water. Cool, and drain upside-down on absorbent kitchen paper. Pack in rigid boxes.	Blanch a few at a time, in a large container for 8–10 minutes. Fonds only 4–5 minutes.
Asparagus	Grade into thick and thin stems but don't tie into bunches yet. Wash in cold water, blanch, cool and drain. Tie into *small* bundles.	Thin stems—2 minutes Thick stems—4 minutes
Aubergines (egg plant)	Peel and cut roughly into 2.5-cm (1-inch) slices. Blanch, chill and dry on absorbent paper. Pack in layers, separated by non-stick paper.	3–4 minutes
Avocados	Prepare in pulp form: peel and mash, allowing 15 ml (1 tbsp) lemon juice to each avocado. Pack in small containers. Also good frozen with cream cheese to use as a party dip.	
Beans French, runner, broad	Select young, tender beans; wash thoroughly French—trim ends and blanch Runner—slice thickly and blanch Broad—shell and blanch In each case, cool, drain and pack.	 2–3 minutes 2 minutes 3 minutes
Beetroot	Choose small beets up to 5-cm (2-inch) diameter. Wash well and rub skin off after blanching. Beetroot under 2.5 cm (1 inch) in diameter may be frozen whole; others should be sliced or diced. Pack in cartons. *Note*: Short blanching and long storage can make beetroot rubbery.	Small whole—5–10 minutes Large—cook until tender—45–50 minutes
Broccoli	Trim off any woody parts and large leaves. Wash in salted water, and cut into small sprigs. Blanch, cool and drain well. Pack in boxes in 1–2 layers, tips to stalks.	Thin stems—3 minutes Medium stems—4 minutes Thick stems—5 minutes
Brussels sprouts	Use small compact heads. Remove outer leaves and wash thoroughly. Make small cuts in stem. Blanch, cool and drain well before packing.	Small—3 minutes Medium—4 minutes
Cabbage green, red	Use only young, crisp, well-hearted cabbage. Wash thoroughly, shred. Blanch, cool and drain. Pack in small quantities in polythene bags.	1–2 minutes
Carrots	Choose small young carrots. If left whole scrape after blanching. Slice or cut into small dice. Blanch, cool, drain and pack.	3–5 minutes

Vegetable	*Preparation*	*Blanching time*
Cauliflower	Heads should be firm, compact and white. Wash, break into small sprigs, about 5 cm (2 inches) in diameter. Add the juice of a lemon to the blanching water to keep them white; blanch, cool, drain and pack.	3 minutes
Celeriac	Wash and trim. Cook until almost tender, peel and slice. Cool then pack.	
Celery	Trim, removing any strings, and scrub well. Cut into 2.5-cm (1-inch) lengths. Suitable only for cooked dishes.	2 minutes
Chestnuts	Wash nuts, cover with water, bring to the boil, drain and peel. Pack in rigid containers. Can be used to supplement raw chestnuts in recipe, can be cooked and frozen as purée for soups and sweets.	1–2 minutes
Chillies	Remove stalks and scoop out the seeds and pithy part. Blanch, cool, drain and pack.	2 minutes
Corn on the cob	Select young yellow kernels, not starchy, over-ripe or shrunken. Remove husks and 'silks'. Blanch, cool and dry. Pack individually in freezer paper or foil. *Note*: There may be loss of flavour and tenderness after freezing. Thaw before cooking.	Small—4 minutes Medium—6 minutes Large—8 minutes
Courgettes	Choose young ones. Wash and cut into 1-cm ($\frac{1}{2}$-inch) slices. Either blanch, or sauté in a little butter.	1 minute
Fennel	Trim and cut into short lengths. Blanch, cool, drain and pack.	2 minutes
Kohlrabi	Use small roots, 5–7 cm (2–3 inches) in diameter. Cut off tops, peel and dice. Blanch, cool, drain and pack.	$1\frac{1}{2}$ minutes
Leeks	Cut off tops and roots; remove coarse outside leaves. Slice into 1-cm ($\frac{1}{2}$-inch) slices and wash well. Sauté for 4 minutes in butter or oil, drain, cool, pack and freeze. Only suitable for casseroles or as a base to vichysoisse.	
Mange-tout (sugar peas)	Trim the ends. Blanch, cool, drain and pack. Use only young tender produce.	2 minutes
Marrow	Young marrows can be peeled, cut into 1–2.5-cm ($\frac{1}{2}$–1-inch) slices and blanched before packing.	3 minutes
Mushrooms	Choose small button mushrooms and leave whole, wipe clean but don't peel or blanch. Sauté in butter for 1 minute. Mushrooms larger than 2.5 cm (1 inch) in diameter: suitable only for slicing and using in cooked dishes.	

Vegetable	*Preparation*	*Blanching time*
Onions	Can be peeled, finely chopped and packed in small plastic containers for cooking later; packages should be overwrapped, to prevent the smell filtering out. *Note*: Small onions may be blanched whole and used later in casseroles.	2 minutes Small whole—4 minutes
Parsnips	Trim and peel young parsnips and cut into narrow strips. Blanch, cool and dry.	2 minutes
Peas, green	Use young, sweet green peas, not old or starchy. Shell and blanch. Shake the blanching basket from time to time to distribute the heat evenly. Cool, drain and pack in polythene bags or rigid containers.	1–2 minutes
Peppers, sweet	Freeze red and green peppers separately. Wash well, remove stems and all traces of seeds and membranes. Can be blanched as halves for stuffed peppers, or in thin slices for stews and casseroles. For better colour, when storage is less than 6 months, do not blanch.	3 minutes
Potatoes	Best frozen in the cooked form, as partially-cooked chips (fully-cooked ones are not satisfactory), croquettes or duchesse potatoes. New: choose small even-sized potatoes. Scrape, cook fully with mint and cool. (Appearance similar to that of canned potatoes.) Chipped: Soak in cold water for about ½ hour. Drain, dry. Part-fry in deep fat for 2 minutes, cool and freeze for final frying.	
Tomatoes	Tomatoes are most useful if frozen as purée. Small whole tomatoes packed in bags can be used in cooked dishes.	
purée	Skin and core tomatoes, simmer in their own juice for 5 minutes until soft. Rub through a nylon sieve or purée in a blender, cool and pack in small containers.	
Turnips	Use small, young turnips. Trim and peel. Cut into small dice, about 1 cm (½ inch). Blanch, cool, drain and pack in rigid containers. *Note*: Turnips may be fully cooked and mashed before freezing—leave 1-cm (½-inch) headspace.	2½ minutes

Unsuitable for freezing: Chicory, cucumber, endive, kale, lettuce, radishes, Jerusalem artichokes (suitable only as soups and purées).

How to freeze fruit

Freezing fruits at their best means freezing them just as they become ready for eating, but if you haven't managed to get around to them in time, slightly over-ripe fruits can still be used for purées. (These take up less room in the freezer, and they're a marvellous stand-by for sauces, desserts and baby foods.) Tough-skinned fruits like redcurrants, blackcurrants and gooseberries or blackberries can be frozen just as they come. Even juicy ones like strawberries and raspberries only need sprinkling with sugar, or freezing free-flow fashion without sugar, though firm-textured fruits like peaches and apricots prefer freezing in a syrup for best results. If they're not to discolour (and this goes for apples and pears, too), you need to soak them in a solution of lemon juice. This increases the food value, and even though fruits may leak some natural sugar and vitamins during thawing, you can eat the fruits in their own syrup and make sure none of them gets wasted. (For syrup, see page 41.)

When using the dry pack method, open freeze the fruit first. Place the fruit on a tray and freeze it until firm, before packing it in containers or bags.

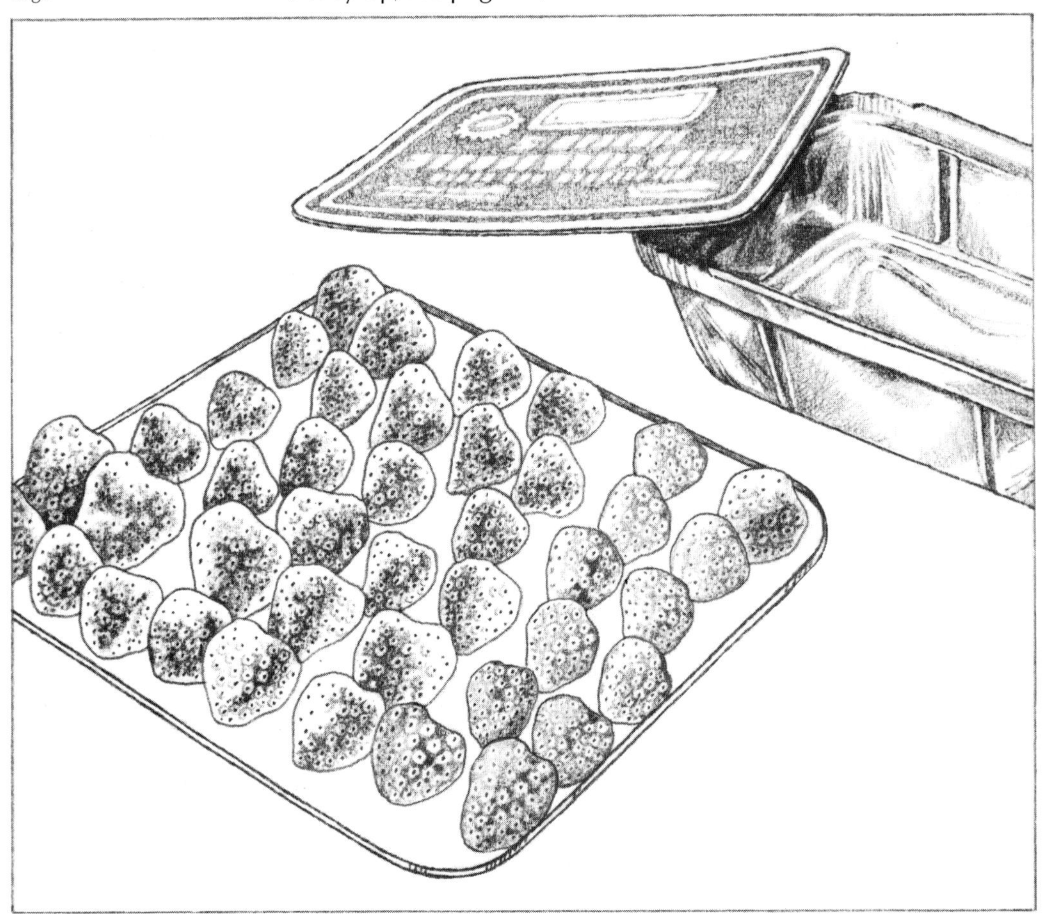

Fruit	*Preparation*
Apples, sliced	Peel, core and drop into cold water. Cut into 0.5-cm ($\frac{1}{4}$-inch) slices. Blanch for 2–3 minutes and cool in ice-cold water before packing. Useful for pies and flans.
purée	Peel, core and stew in the minimum amount of water—sweetened or unsweetened. Purée or mash. Leave to cool before packing.
Apricots	Plunge them into boiling water for 30 seconds to loosen the skins, then peel. Then prepare in one of the following ways: (a) Cut in half or slice into syrup made with 450 g (1 lb) sugar to 1 litre (2 pints) water, with some lemon juice added to prevent browning; for each 450-g (1-lb) pack allow the juice of 1 lemon. Immerse the apricots by placing a piece of clean, crumpled, non-absorbent paper on the fruit, under the lid. (b) Leave whole, and freeze in cold syrup. After long storage, an almond flavour may develop round the stone. (c) Purée cooked apricots. Pack sweetened or unsweetened.
Berries (including currants and cherries)	All may be frozen by the dry pack method, but the dry sugar pack method is suitable for soft fruits, eg raspberries. *Dry Pack:* Sort the fruit; some whole berries may be left on their sprigs or stems for use as decoration. Spread the fruit on paper-lined trays or baking sheets, put into the freezer until frozen, then pack. *Dry Sugar Pack:* Pack dried whole fruit with the stated quantity of sugar: 100–150 g (4–6 oz) to 450 g (1 lb) fruit. Mix together and seal.
Blackberries	Dry pack or dry sugar pack—allow 225 g (8 oz) sugar to 900 g (2 lb) fruit. Pack in rigid containers.
Blueberries or Bilberries	Wash in chilled water and drain thoroughly. Use one of the following methods: (a) Dry pack. (b) Dry sugar packed—about 100 g (4 oz) sugar to 450–700 g (1–1$\frac{1}{2}$ lb) fruit; slightly crush berries, mix with sugar until dissolved and then pack in rigid containers. (c) Frozen in cold syrup—900 g (2 lb) sugar dissolved in 1 litre (2 pints) water.
Gooseberries	Wash and thoroughly dry fruit. Use one of the following methods: (a) Dry method in polythene bags, without sugar; use for pie fillings. (b) In cold syrup—900 g (2 lb) sugar to 1 litre (2 pints) water. (c) Purée—stew fruit in a very little water, rub through a nylon sieve and sweeten to taste; useful for fools and mousses.
Loganberries	Choose firm, clean fruit. Remove stalks and dry pack in rigid containers. Dry sugar pack—see Blackberries.
Strawberries and Raspberries	Choose firm, clean, dry fruit; remove stalks. Raspberries freeze well. Whole strawberries can be a disappointment. Add a little lemon juice to strawberry purée. Use one of the following methods: (a) Dry method. (b) Dry sugar pack—100 g (4 oz) sugar to each 450 g (1 lb) fruit. (c) Purée; sweeten to taste—about 50 g (2 oz) sugar per 225 g (8 oz) purée—and freeze in small containers; useful for ice creams, sorbets, sauces or mousses.
Blackcurrants	Wash, dry, top and tail. (a) Use dry pack method for whole fruit. (b) Purée—cook to a purée with very little water and brown sugar, according to taste.
Redcurrants	Wash and dry the whole fruit, then use dry pack method. Pack in rigid containers.

Fruit	*Preparation*
Cherries	Remove the stalks. Wash and dry. Use one of the following methods: (a) Dry pack. (b) Dry sugar pack 225 g (8 oz) sugar to 900 g (2 lb) stoned cherries; pack in containers cooked or uncooked, best used stewed for pie fillings. (c) Cover with cold syrup 450 g (1 lb) sugar to 1 litre (2 pints) water, mixed with 2.5 ml ($\frac{1}{2}$ tsp) ascorbic acid per 1 litre (2 pints) syrup; leave headspace. Take care not to open packet until required, as fruit loses colour rapidly on exposure to the air.
Damsons	Wash in cold water. The skins are inclined to toughen during freezing. Best packing methods are: (a) Purée. (b) Halve, remove the stones and pack in cold syrup—450 g (1 lb) sugar to 1 litre (2 pints) water; they will need cooking after freezing—can be used as stewed fruit. (c) Poached and sweetened.
Figs	Wash gently to avoid bruising. Remove stems. Use one of the following methods: (a) Freeze unsweetened, either whole or peeled, in polythene bags. (b) Peel and pack in cold syrup—450 g (1 lb) sugar to 1 litre (2 pints) water. (c) Leave whole and wrap in foil—suitable for dessert figs.
Grapefruit	Peel fruit, removing all pith; segment and pack: (a) In cold syrup (equal quantities of sugar and water—use any juice from the fruit to make up the syrup). (b) Dry sugar pack—225 g (8 oz) sugar to 450 g (1 lb) fruit, sprinkled over fruit; when juices start to run, pack in rigid containers.
Grapes	The seedless variety can be packed whole; others should be skinned, pipped and halved. Pack in cold syrup—450 g (1 lb) sugar to 1 litre (2 pints) water.
Greengages	Wash in cold water, halve, remove stones and pack in syrup—450 g (1 lb) sugar to 1 litre (2 pints) water, with the juice of 1 lemon added. Place in rigid containers. Do not open pack until required, as fruit loses colour rapidly. Skins tend to toughen during freezing.
Lemons and Limes	Use one of the following methods: (a) Squeeze out juice and freeze it in ice-cube trays; remove frozen cubes to polythene bags for storage. (b) Leave whole, slice or segment before freezing. (c) Remove all pith from the peel, cut into julienne strips, blanch for 1 minute, cool and pack; use for garnishing dishes. (d) Mix grated lemon peel and a little sugar to serve alongside pancakes. (e) Remove slivers of peel, free of pith, and freeze in foil packs to add to drinks.
Mangoes	Peel and slice ripe fruit into cold syrup—450 g (1 lb) sugar to 1 litre (2 pints) water; add 30 ml (2 tbsp) lemon juice to each 1 litre (2 pints) syrup. Serve with additional lemon juice.
Melons	Cantaloup and honeydew melons freeze quite well (though they lose their crispness when thawed), but the seeds of watermelon make it more difficult to prepare. (a) Cut in half and seed, then cut flesh into balls, cubes or slices and put straight into cold syrup—450 g (1 lb) sugar to 1 litre (2 pints) water. (b) Dry pack method, with a little sugar sprinkled over. Pack in polythene bags.

Fruit	*Preparation*
Oranges	Prepare and pack as for grapefruit, or use one of the following methods: (a) Squeeze out and freeze the juice; add sugar if desired and freeze in small quantities in containers or in ice-cube trays. (b) Grate peel for orange sugar as for lemon sugar. (c) Seville oranges may be scrubbed, packed in suitable quantities and frozen whole until required for making marmalade. (Do not thaw whole frozen fruit in order to cut it up before cooking as some discoloration often occurs—use whole fruit method for marmalade. It is advisable to add one-eighth extra weight of Seville or bitter oranges or tangerines when freezing for subsequent marmalade making in order to offset pectin loss.)
Peaches	Really ripe peaches are best skinned and stoned under running water, as scalding will soften and slightly discolour the flesh. Treat firm peaches in the usual way. Brush over with lemon juice. (a) Pack halves or slices in cold syrup—450 g (1 lb) sugar to 1 litre (2 pints) water, with the juice of 1 lemon added; pack in rigid containers, leaving 1-cm ($\frac{1}{2}$-inch) headspace. (b) Purée peeled and stoned peaches; mix in 15 ml (1 tbsp) lemon juice and 100 g (4 oz) sugar to each 450 g (1 lb) fruit—suitable for sorbets and soufflé-type desserts.
Pears	It is really only worthwhile freezing pears if you have a big crop from your garden, as they discolour rapidly, and the texture of thawed pears can be unattractively soft. Peel, quarter, remove core and dip in lemon juice immediately. Poach in syrup—450 g (1 lb) sugar to 1 litre (2 pints) water—for $1\frac{1}{2}$ minutes. Drain, cool and pack in the cold syrup.
Pineapple	Only freeze quality *ripe* fruit. Peel and core, then slice, dice, crush or cut into wedges. (a) Pack unsweetened in boxes, separated by non-stick paper. (b) In cold syrup—450 g (1 lb) sugar to 1 litre (2 pints) water—in rigid containers; include any pineapple juice from the preparation. (c) Pack the crushed pineapple in rigid containers, allowing 100 g (4 oz) sugar to about 350 g ($\frac{3}{4}$ lb) fruit.
Plums	Wash, halve and discard stones. Freeze in syrup; use 450 g (1 lb) sugar to 1 litre (2 pints) water with the juice of 1 lemon. Pack in rigid containers. Do not open packet until required, as the fruit loses colour rapidly.
Rhubarb	Wash, trim and cut into 1–2.5-cm ($\frac{1}{2}$–1-inch) lengths. Heat in boiling water for 1 minute and cool quickly. Pack in cold syrup, using equal quantities sugar and water, or dry pack, to be used later for pies and crumbles.

Fruits not suitable for freezing: bananas, pomegranates.

The syrup

When preparing fruits for the freezer, make sure you have sufficient syrup available. This should be made in advance, and the best plan is to make it a day ahead and allow it to chill overnight in the refrigerator, since it has to be used cold. As a rough guide, for every 450 g (1 lb) fruit you need 300 ml ($\frac{1}{2}$ pint) syrup (made according to the individual directions). Dissolve the sugar in the water, bring to the boil, remove from the heat, strain if necessary, add lemon juice where indicated, and leave to cool, lightly covered. Pour the syrup over the fruit or place the fruit in a container with the syrup. Light-weight fruits which tend to rise in liquids can be held below the surface by means of a dampened and crumpled piece of non-absorbent paper.

Incidentally, when you come to serve fruit frozen in syrup, only open it just before serving; keep stoned fruit submerged in the syrup as long as possible.

How to freeze meat

The object of making significant financial saving by buying carcass meat and freezing at home for later consumption is rarely achieved. In the first place a very large freezer would be required to freeze and store the various cuts of meat but also you need to utilise all the animal, so consider family likes and dislikes. It may therefore be more economical to buy ready-butchered frozen meat in the joints that are used regularly. Bought in bulk a saving can be achieved.

Already-butchered fresh meat demands next to no preparation. Excess fat should always be removed, because it tends to go rancid quicker than the flesh during storage, and whenever possible, it makes sense to remove any bones as well. These only take up freezer space without giving any return for your money, and you can always use them for making soups and stocks, which are much more economical to store.

Tackling a carcass

Butchering is a skilled job which takes years to learn. The difficulties of coping with a carcass are many. First, you'll need an enormous freezer. A mere hind or forequarter of beef, for instance, is going to weigh about 68 kg (150 lb)—enough to half-fill a 340-litre (12 cubic-foot) freezer. And as your family's going to get heartily sick of beef unless you've frozen some pork, lamb, veal and poultry as well, there'll be precious little room left for equally important items such as fruit, vegetables and pre-cooked dishes.

That same hind of beef is only a good buy when greatly reduced in price. It might not be quite such a bargain when you've had to throw away (or process) 13–18 kg (30–40 lb) of it in bones, fat and waste. Nor does a carcass consist of all steaks and succulent joints, but things like shin and flank as well.

If the carcass is freshly slaughtered, you're going to have to chill it, in the case of pork and veal; also hang it, in the case of beef and lamb—a tedious and long-drawn-out process. Once you've chilled pork or veal, it has to be frozen *immediately* if it's to have a long storage life.

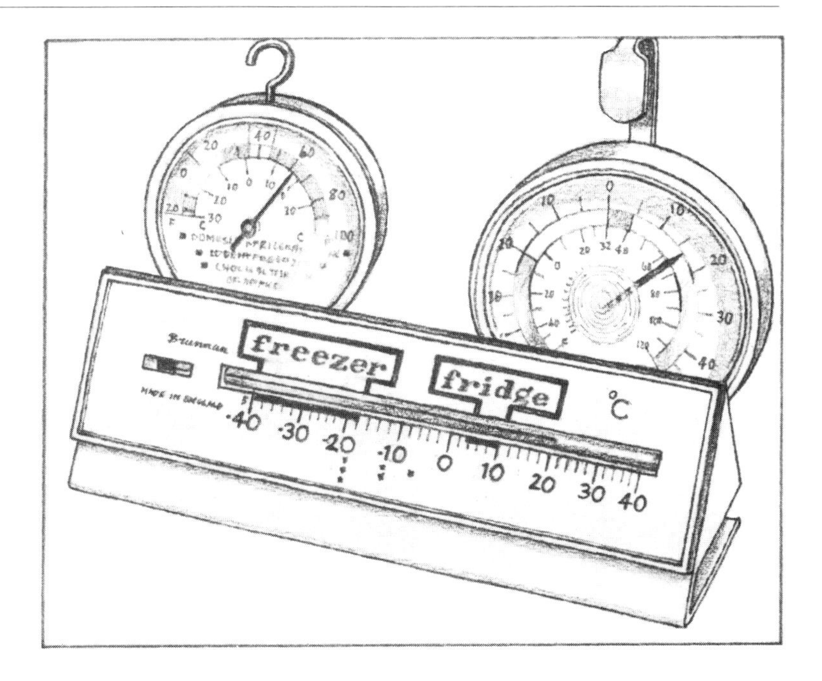

A freeze thermometer will help you to monitor the efficiency of your freezer.

Probably the largest amount of carcass-meat within the average person's freezing scope is a forequarter of lamb. If it's freshly slaughtered, first chill it in the refrigerator at 5°C (40°F) as quickly as possible; something that will take between 1–2 days. Then hang it, which will take between 6–12 days. Good hanging—always important for flavour and texture—is especially vital when meat is to be frozen, as it reduces the amount of drip when the meat is thawed. Next, invest in a tiny saw and a really sharp boning knife, and cut the carcass into neat joints in the following order: the shoulder; the breast; the middle and best end (these should give you five prime cutlets and three middle neck, but switch from your knife to your saw and chine them before freezing in total; saw right through them if you want to freeze cutlets separately); trim off the neck tendon from the stewing end, and bone the breast. Run up either side of the cartilage to remove cartilage bone, having taken the flap back, and remove the bones, firm skin and excess fat before freezing.

That's enough to prove it's a skilled job, and if you do get access to larger amounts of carcass going cheap, it's usually worth spending a little of what you're saving on a butcher's services. You may have to ask around until you find one who's willing, but it's far more satisfactory to have it chilled, hung and jointed for you—even frozen in commercial blast-freezing facilities, which do produce the best results for meat. This is because meat needs to be frozen quickly, and unless you have a fast-freeze switch to get your freezer temperature down to −24°C (−11°F) or below and keep it there, you might only be capable of producing slow-frozen results.

Of course, if you're offered sound-quality meat at a *very* low price, it may be worth while buying it to freeze at home, even though the end result may not be quite so excellent as the blast-frozen product.

If you are freezing your own meat, remember that very lean meats tend to dry out. A thin layer of fat and good 'marbling' help to prevent this; however, too much fat tends to become rancid eventually. This is why excess fat is best removed wherever practical and—as the speed at which fat goes rancid depends on the amount of pre-freezing oxygen present—removal of air and good packaging are vital. This is not so vital when considering short-term storage. A freezer thermometer is a worthwhile investment if you're freezing large quantities of meat.

Food	*Recommended storage time*	*Food*	*Recommended storage time*
beef	8 months	casseroles	
lamb	6 months	with bacon	2 months
mince	3 months	without bacon	3 months
offal	3 months	curry	4 months
pork	6 months	ham	2 months
sausages	3 months	meat pies	3 months
veal	6 months	pâté	1 month
		sliced meat	
		with gravy	3 months
		without gravy	2 months
		shepherd's pie	3 months
		soup	3 months
		stock	6 months

How to freeze poultry

Freezing chickens (including capons and poussins)

It's not worth spending your efforts on anything other than young, plump, tender birds. Commercial quick-frozen raw poultry is so readily available it's only an advantage to freeze it at home when the price is very favourable.

If the birds are your own, starve them for 24 hours before killing, and if possible set to and pluck out the feathers while the corpse is still warm—they come out much easier that way.

Hang the chicken for a day before drawing, and be sure to freeze at the lowest possible temperature for your freezer—ideally, −32°C (−26°F)—because higher temperatures can give disappointing results. And remember to turn the temperature control down low at least 24 hours before you're planning to freeze a bird—this will give it a chance to get really cold. Once the chicken is frozen, return to the normal freezer temperature.

How you freeze your birds depends on how you're going to cook them. Obviously if you're going to make a chicken casserole, you'll be wasting freezer space if you freeze a chicken whole, instead of cutting it up into more convenient pieces.

And talking of wastage, remember that chicken fat is excellent rendered down for cooking chips in, and that any odd scraps and bones can make very good concentrated stocks for using in soups and sauces.

More sizeable remnants of cold roast chicken (carved or not) and poached chicken may be frozen satisfactorily too.

WHOLE

Wipe and dry the bird after plucking and drawing it. Pack the giblets separately instead of inside, because they'll only keep for a quarter of the time the chicken can be stored. Don't stuff the chicken, as it takes too long to freeze and thaw. If you wish, package any stuffing separately.

To truss the bird, place it on the table with the breast uppermost and the legs on the right-hand side ready for trussing. Insert the trussing needle, threaded with fine string, through the top joint of one leg, through the body and out through the other leg, leaving an end of string. Catch in the wing, then pass the needle through the body and catch in the bottom of the opposite leg. Again, insert the needle through the bottom end of the leg, pass the needle through the body and through the bottom of the opposite leg. Finally, pass the needle diagonally through the body and catch in the remaining wing. Tie the string tightly.

To truss the bird without using a special needle, insert a skewer right through the body just below the thigh bone and turn the chicken over on its breast. Catching in the wing tips, pass the string under the ends of the skewer and cross it over the back. Turn the bird over and tie the ends of the string round the tail, at the same time securing the drum sticks.

PORTIONS

Divide small birds, around 1–1.5 kg (2–3 lb), into quarters. With young birds, this can be done with poultry secateurs or a sharp knife. Cut the bird in half, through and along the breastbone. Open the bird out, and then cut along the length of the backbone.

If you want to remove the backbone entirely, cut along either side of it, then lift it out—and don't forget to use it for making stock. If you're using a knife, you'll have to tap the back sharply with a heavy weight to cut through the bony sections.

Either way, once the bird is in two halves, lay these skin side up. Divide each in half again by cutting diagonally across between wing and thigh—allocating more breast meat to the wing than to the thigh, to even out the amount of meat per portion.

SMALLER CHICKEN JOINTS

When these are to be used in casseroles, cut the thigh loose along the rounded edge and pull the leg away from the body to isolate the joint. Break the thigh backwards so that the knife can cut through the socket of each thigh joint, and loosen the wings from the breast meat in the same way. Divide the legs into two pieces in the centre of the joint. (Use large birds for jointing.)

Lastly, turn over the body of the bird on to its back and carve the breast meat from the breastbone. Both breast portions may be halved, and the back is divided into two or three pieces or used to supplement the stock pot.

It is worthwhile dividing a chicken into portions for freezing. Portions will thaw more rapidly than a whole bird and are fine for casseroles.

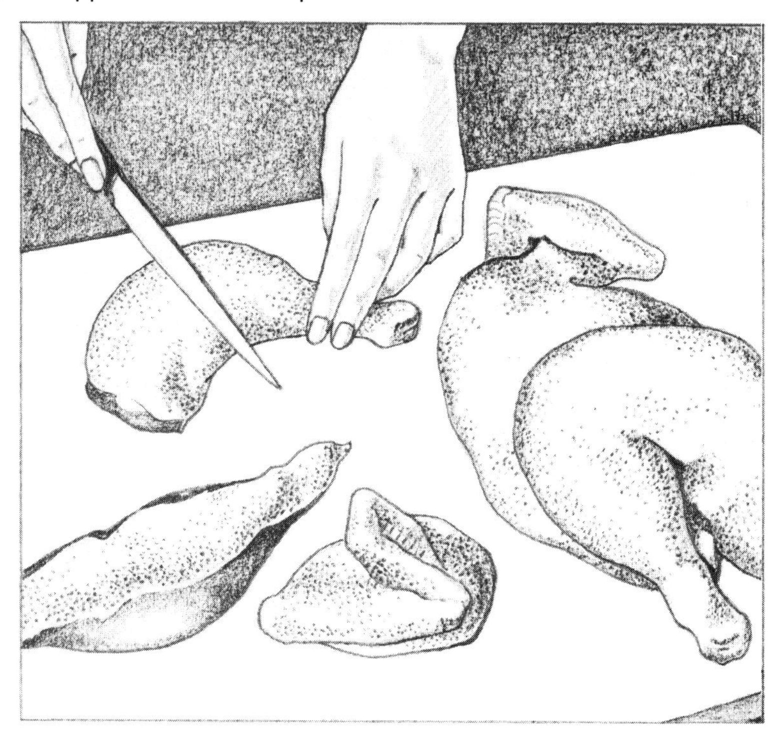

PACKING FOR THE FREEZER

Before packing whole birds in freezer bags, first pad the legs with foil so that they can't spike their way through the wrapping. Exclude as much air as possible before sealing the bag (see pages 23 to 24).

With chicken quarters or joints, pack individually in foil or polythene bags—don't forget to label them—and then combine into a larger package to save time hunting for individual packs.

Cold roast or poached chicken should be cooled as rapidly as possible after cooking. Parcel small amounts in foil, with any stuffing packed separately, and freeze at once.

The recommended storage time for chicken is 12 months, and for giblets 2–3 months. Cold cuts 2–3 months.

Turkeys

In general, these are treated like chickens. But because they are so bulky, they take up a lot of valuable freezer space, so it's not good planning to store them for too long. The maximum storage time is 6 months.

Leftover roast turkey may be frozen as for chicken. It's best cut off the bone, and any fat should be discarded, while stuffing should be packed separately. To avoid excessive drying, freeze it with gravy or stock, unless it's only to be stored for a very short time. Storage time 2–3 months.

Ducklings and geese

Choose young birds without too much fat, which can cause rancidity—although excessively lean birds may be dry. Dress in the usual way, packing the giblets separately and freeze as for chicken.

Roasting

All frozen poultry must be completely thawed before cooking. (See page 55 for approximate thawing times.)

Weigh bird after stuffing. Set oven at 180°C (350°F) mark 4. Place bird in roasting tin (or rack if available). Cover with foil, wrapping it over the edge of the pan, tent fashion. Roast according to the chart below. To brown, remove foil for last 30 minutes.

Approximate Roasting Times (using foil) for Thawed Poultry

Weight	*Roasting time*
900 g (2 lb)	1¾ hours
1.4 kg (3 lb)	2 hours
1.8 kg (4 lb)	2¼ hours
2.3 kg (5 lb)	2½ hours
4.5 kg (10 lb)	3½ hours
6.8 kg (15 lb)	4¾ hours

How to freeze game

Venison

Like ordinary meats, venison requires some skill in the art of butchery, so it's best to ask the butcher to deal with it, if you're ever confronted with a carcass. Briefly, it needs to be chilled, hung for up to 8 days, then jointed and frozen, packed and thawed like other meats. Cook as for fresh venison. It has a storage life of up to 12 months.

Hares and Rabbits

Hares are game, rabbits aren't—but it's convenient to bracket them together here. Prepare them as for cooking fresh. If you like the gamey flavour, hang a hare for 5–7 days before freezing it, but rabbits are not hung.

Since most recipes call for portioned hare or rabbit, it's sensible to pack them this way, discarding the more bony parts (which can be used for casseroles, pâté and stock). Pack and freeze as for meat.

The recommended storage time is 6 months.

Game birds

These actually improve with freezing. Bleed the bird as soon as possible after shooting. Hang, undrawn, until sufficiently 'high' for your personal taste. Five–6 days is long enough before freezing. If necessary, a bird can be frozen for a *short* time with its feathers on, and plucked after thawing.

Pluck, removing the feathers in the direction in which they grow. Draw, wash, drain and thoroughly dry. Pack, removing as much air as possible, in a really heavy-gauge foil or freezer bag before freezing in the usual way.

The recommended storage time is 9 months.

Water birds

These should be plucked, drawn and frozen quickly after killing. (Remove the oil sac from the base of the tail.)

The recommended storage time is 6 months.

How to freeze bacon

Bacon can be a useful freezer item, but it is very important to follow directions carefully and not to exceed the recommended storage times, otherwise both texture and flavour may be disappointing.

It is most important to freeze bacon which is perfectly fresh. If possible, it should be frozen on the day it is supplied to the retailer —the longer the bacon has been cut or kept in the shop, the shorter its storage life in the freezer.

Smoked bacon can be stored for longer periods than unsmoked before any risk of rancidity arises.

To pack and freeze

Commercial vacuum-packing is generally accepted as the most suitable wrapping, since the maximum amount of air is extracted, so if possible buy bacon for freezing ready-packed in this way. Again, the bacon should be frozen well before the sell-by date.

However, quite acceptable results have been achieved when top-quality bacon has been closely wrapped in freezer film or foil and overwrapped in polythene bags. Don't have more than 225 g ($\frac{1}{2}$ lb) rashers in a package. If you wish, they can be interleaved for easier separation—use waxed or non-stick paper or Cellophane—

Right for any occasion: two succulent casseroles—Chicken Casserole garnished with bacon and baby onions, and Chicken and Pepper Fricassée—and delicious Devilled Drumsticks.

Overleaf: Savour the taste of Osso Buco with Tomato from sunny Italy, or try Lamb with Apricots, a favourite Arabic dish, or the unmistakably French Pork and Bacon Casserole with pastry fleurons.

but this is not necessary if the packs are made up to suit individual cooking needs.

Wrap bacon chops individually in foil, then pack together in a polythene bag. Joints up to 1.5–2 kg (3–4 lb) should be wrapped in foil then overwrapped in a polythene bag. It is also wise to overwrap vacuum-packed bacon, and to check each one carefully to make sure the vacuum is intact and that the bacon does not move about within the pack. Freeze as quickly as possible.

Food	*Recommended storage times*
Vacuum-packed bacon	3 months
Smoked rashers, chops, gammon steaks and joints	2 months
Unsmoked rashers, chops, gammon steaks and joints	1 month

Storage

Once the food is in the freezer, how long can you leave it there? As far as safety goes, practically for ever as long as you maintain a constant temperature of −18°C (0°F). Food-poisoning bacteria cannot multiply in or on frozen food, so there is no danger of the food becoming a health hazard, no matter how long it is stored. However, food contaminated before it has been frozen will still be contaminated after it has been thawed. As with all perishable foods, if frozen food is allowed to thaw out and is kept at room temperature (unrefrigerated), bacteria will develop and it will become a health hazard.

As far as food value, taste, colour and texture go, different foods have different storage times, which are determined by the length of time foods can be stored without any detectable change in eating quality. Since each person's ability to detect these changes varies considerably there will be differing opinions about how long to store food.

Maximum recommended storage time for most vegetables is 10–12 months (only 6–8 for vegetable purées); for most fruits packed in syrup or sugar, 9–12 months (but only 6–8 for fruits packed alone or as purées—and only 4–6 months for fruit juices). Meats like beef and lamb can be stored for up to 8 months, but fattier meats like pork should preferably be kept for no longer than 6 months. This is because the fat tends to oxidise and go rancid, something that destroys fat-soluble vitamins such as vitamin A, as well as spoiling flavour. Oxidation can take place in oily fish like mackerel, too, which is why they can only be kept for 2 months, as opposed to 3 months for white fish.

All these estimates ('Basic Know-how' on pages 57 to 64 provides more specific information) are on the conservative side,

These tempting cakes and cookies—Layered Chocolate Slices, Ginger Biscuits, Almond Bars and Almond Small Cakes—are just some of the fruits of a day's baking.

because, although the ideal is to store food at a *constant* temperature of −18°C (0°F), this is a counsel of perfection. Every time you take something out of the freezer, the temperature is going to rise, and it's going to take some time for the thermostat-controlled compressor to bring it back to normal. Similarly, every time you add an unfrozen product, you're going to create a small temperature gradient—though you can keep it to the minimum by chilling packages in the refrigerator first.

These small fluctuations of temperature are fairly harmless. Big swings aren't, however, because even if you've quick-frozen your food, any pronounced fluctuation is still going to affect ice crystal formation—with possible loss of quality.

Really drastic rises in temperature excluded and presuming your food's been efficiently packaged, it shouldn't come to any real harm through faulty storage. Even if a temperature gradient has been set up, causing air to circulate and draw out more moisture than usual, your food won't be able to suffer desiccation or freezer burn if none of the air can reach it. In fact, the only scope for damage inside an airtight package is if you didn't manage to squeeze out all the air to begin with. Then, it's possible for air to circulate *within* the package, drying out the food and depositing the moisture on the insides of the pack as cavity ice—a process that can't be put right at the thawing stage. All of which underlines just how vital the packaging stage is.

Thawing

There are quite a few foods that need no thawing when they come out of the freezer.

Stews, pre-cooked pies and casseroles can go into the oven still frozen, but unless you freeze them in fairly shallow dishes to begin with, there's always the danger they won't have reheated right through. Ideally thaw at cool room temperature. As well as giving good reheating results it is more economical of fuel.

Meat Small cuts of meat like chops, steaks and mince can also be gently cooked straight from frozen, if time does not allow for thawing and so can some whole joints of meat (see page 53), provided you can be sure the inside cooks before the outside overcooks.

Bacon must be thawed before cooking.

Sausages tend to burst if cooked from frozen.

Vegetables *shouldn't* be thawed (and if you're a real nutrition-fiend, why not try slow-cooking them in a heavy saucepan with a knob of butter instead of water, so that you can make the most of every vitamin going). One exception is corn on the cob.

Bread and rolls Thawing them in a warm oven crisps them up, and makes them as fragrant as if you'd just baked them.

Poultry *must be completely thawed out* prior to cooking.

Fish Whole fish should be thawed out as slowly as possible if they're to retain their juiciness and texture. Small fish and cuts are best cooked from frozen.

Fruit that's to be eaten without further preparation needs the gentlest possible thawing. This minimises the amount of leakage (not that this will bother you in terms of food value if you're eating it in its own syrup), and prevents it from going too soft and mushy. Even with the most considerate thawing, though, fruit tends to 'fall', which is why many people eat it while there's still a little ice in it.

SLOW THAWING

The slowest way of thawing is in the refrigerator or at cool room temperature. Always leave food in its original wrapping, because otherwise meat will bleed and lose some of its quality and colour; fruit will lose its juice and suffer in texture; fish will dry out and generally coarsen. And always leave yourself plenty of time. As little as 450 g (1 lb) of food can take anything up to 6 hours to thaw, which means a 1.8 kg (4 lb) salmon trout could keep you waiting for 24 hours.

QUICK THAWING

Many people sacrifice a little in the way of quality for quicker results. Food thawed at normal room temperature only takes about half the time—and in real emergencies, you can submerge the package in a bowl of cold water or hold it under a running tap.

STORING THAWED FOOD

However you thaw the food, don't leave it lying around at normal room temperature for longer than necessary. Although freezing brings micro-organisms to a grinding halt, remember that they try to make up for lost time as soon as thawing sets them free. This means that food becomes extra-perishable once it's thawed—so put it into the fridge, or preferably, eat it right away.

Is it safe to re-freeze thawed food?

Partly-thawed food—provided that the packet hasn't been opened—may be refrozen without danger to health. But the eating quality of the food deteriorates, so it is not a practice to be recommended.

Cooking and reheating from frozen

In general, frozen foods you want to serve hot should be heated as rapidly as possible, as this preserves the flavour and texture. Whenever practical, heat from frozen. If you're using the oven, you'll find that a temperature of about 200°C (400°F) mark 6 will be fine for combined thawing and reheating from frozen.

As a guide, a shallow 1-litre (2-pint) pack will take about 1 hour. (The advantage of freezing in smallish packs is obvious when it comes to heating up.) You can speed up the process a little by leaving the casserole or what-have-you unlidded. As the food thaws, fork it over gently to separate the pieces. A sauce, once thawed, should be vigorously whisked to restore its smoothness, unless this would break down any firm pieces.

If you have only a small amount of food to reheat, you can do it very satisfactorily in a double boiler; if you are in a hurry, a mixture that includes some sauce or liquid can be put into a saucepan placed directly over low heat, but beware of sticking.

Cooking meat from frozen

The quickest way to thaw meat is to cook it from frozen. Boned and rolled cuts of meat should however not be cooked from frozen. Cooking times for frozen joints over 2.8 kg (6 lb) are very difficult to calculate. To prevent the outside being overcooked before the inside is thawed, it is better to thaw large joints before cooking.

To roast from frozen successfully, it is essential to use a meat thermometer. To test if cooked, shortly before the estimated cooking time, insert the thermometer into the centre of the meat or, if applicable, as near the bone as possible, making sure that the point of the thermometer does not touch the bone.

BONED AND ROLLED JOINTS

Rolled joints such as breast of lamb, whether stuffed or not, must be thawed before cooking. This is because all the surfaces of the meat have been handled and rolled up, so it is important to ensure thorough cooking to destroy any bacteria which might be present in the meat.

SMALL CUTS OF MEAT

Chops, steaks, liver slices and kidneys can be easily cooked from frozen and so can cubed meat for stews and kebabs. Start the cooking at a lower temperature than normal and cook for almost twice as long, increasing the temperature half way through the cooking. It will probably be necessary to thaw meat for frying if it has become mis-shapen during frozen storage to ensure good contact with the frying pan.

POT ROASTING AND BOILING

For the best results, it is advisable to thaw pot-roasting and boiling joints before cooking.

STEWING

Cubed and minced meat can be cooked from frozen. Allow extra time for cooking to allow for thawing.

When cooking a joint from frozen it is essential to use a meat thermometer.

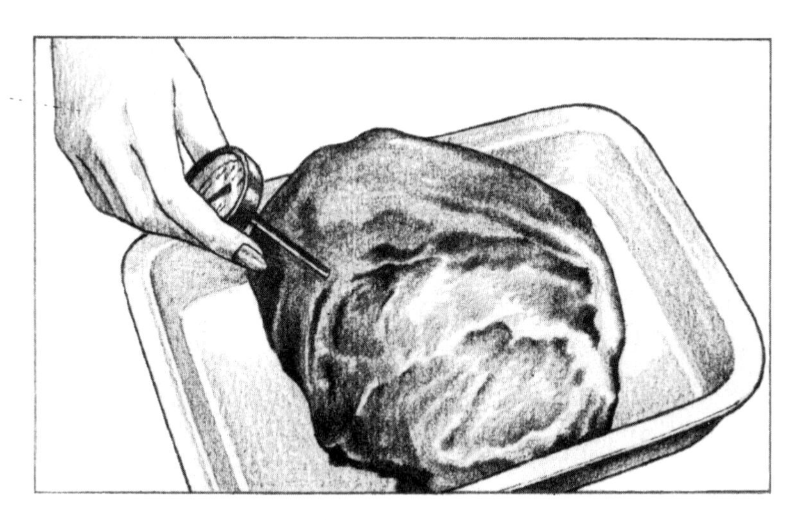

Roasting Meat from Frozen

Meat joints on the bone	*Approximate cooking time at 180°C (350°F) mark 4*	*Thermometer*
Beef		
under 1.8 kg (4 lb)	Well done—35 minutes per 450 g (1 lb) + 35 minutes.	79°C (170°F)
	Medium/rare—30 minutes per 450 g (1 lb) + 30 minutes.	71°C (160°F)
1.8–2.8 kg (4–6 lb)	Well done—40 minutes per 450 g (1 lb) + 40 minutes.	79°C (170°F)
	Medium/rare—35 minutes per 450 g (1 lb) + 35 minutes.	71°C (160°F)
Lamb		
under 1.8 kg (4 lb)	Well done—35 minutes per 450 g (1 lb) + 35 minutes.	82°C (180°F)
1.8–2.8 kg (4–6 lb)	Well done—40 minutes per 450 g (1 lb) + 40 minutes.	
Pork		
1.8–2.8 kg (4–6 lb)	Well done—45 minutes per 450 g (1 lb) + 45 minutes.	88°C (190°F)

Note: Continue cooking until correct temperature is reached. If a roasting bag is used, reduce cooking temperature by 15 minutes per 450 g (1 lb).

Cooking Sausages and Liver From Frozen		
Meat	*Approximate time*	*Further instructions*
Large sausages		
baked	1 hour	At 180°C (350°F) mark 4.
grilled	20 minutes	Start at a low temperature. Turn occasionally during cooking. Tend to split when fried.
Small chipolata sausages		
baked	45 minutes	At 180°C (350°F) mark 4.
grilled	15 minutes	
Sliced liver		
fried	15–20 minutes depending on thickness of cut.	

Thawing and freezing meat

It is perfectly safe to cook thawed meat and then refreeze it, provided it is cooled quickly before freezing. This will be necessary anyway, if casseroles are to be made from bulk-purchased frozen meat. It's as well to cover the cooked meat with well-skimmed sauce, before refreezing, to prevent the meat drying in the freezer.

In all cases it is recommended that meat should be thawed in the refrigerator. Although it will take longer to thaw than it would at room temperature, there is less risk of contamination in the refrigerator. If the juices which have come from the meat during thawing are used in cooking, they should be cooked and eaten, as soon as possible.

Approximate Thawing Times		
Type of cut	*In a refrigerator*	*At room temperature*
Joints 1.4 kg (3 lb) or more	6–7 hours per 450 g (1 lb)	2–3 hours per 450 g (1 lb)
less than 1.4 kg (3 lb)	3–4 hours per 450 g (1 lb)	1–2 hours per 450 g (1 lb)
Steaks, chops, stewing steak (not cubed)	5–6 hours	2–4 hours

Poultry

All frozen poultry must be completely thawed before cooking, so leave plenty of time. We recommend thawing in the refrigerator for poultry up to 2.7 kg (6 lb) and over that at room temperature. Remove giblets and neck as soon as they are free (if they are inside the bird) and use them for making stock or gravy. Remember that once thawed, poultry should be cooked as soon as possible.

Approximate Thawing Times

Weight	*Oven-ready*	
1.8 kg (4 lb)	12 hours	in the refrigerator
2.7 kg (6 lb)	15 hours	
4.5 kg (10 lb)	18 hours	at room temperature
6.8 kg (15 lb)	24 hours	
9 kg (20 lb)	30 hour	

Chickens To keep in all the juices, remember to thaw chicken while it is still in its packing material. Chicken portions need only about 3 hours to thaw.

Don't be alarmed if the poultry bones have turned a sinister darkish-red. It's all to do with oxidation of haemoglobin, and is nothing to worry about.

To thaw whole roast chicken, leave it loosely wrapped at room temperature for about 9 hours. Previously carved meat will thaw more quickly, but may be a little dried in texture.

Turkeys, ducklings and geese, like chickens, should be thawed in their wrappings.

Game

Venison Thaw like other meats. Cook as for fresh venison.

Hares and rabbits Thaw as for meat, and cook as usual.

Game birds Thaw thoroughly, preferably in the refrigerator, and cook immediately.

Bacon

Bacon joints Allow bacon plenty of time to thaw slowly in a cool place. This is important because the more slowly the meat is thawed the better the quality will be when cooked. Thawing in the refrigerator is recommended because the cool temperature will continue to preserve the maximum freshness of the meat during the time it takes for the inside fibres to thaw completely.

The wrapping material should be opened—or the end cut off in the case of vacuum packs—on removal from the freezer. Remove wrappings completely as soon as possible during the thawing process. Joints may be thawed in running cold water, but wrap well in a plastic bag first to prevent them getting wet.

Bacon rashers and small pieces Packets of bacon rashers, steaks, chops may be thawed overnight in the refrigerator or, if wanted for immediate use, can be thawed in hot water for a few minutes. Dry them well on kitchen paper before frying or grilling. Bacon rashers which have been frozen tend to be rather less moist than fresh raw bacon, and consequently they can taste rather salty. To overcome this, dip the rashers in hot water and dry them on kitchen paper before cooking.

Once cooked the bacon will keep as long as other bacon which has not been frozen.

Do not refreeze raw bacon after thawing.

Cooked bacon in rechauffé dishes Small pieces of cold cooked bacon may be used in stews or such dishes as quiche lorraine which are to be put in the freezer for short periods of up to 6 weeks.

Know-how charts for the freezer cook

Even the most accomplished freezer owner cannot have all the freezer facts at fingertips. But basic know-how is the key to success in anything, and freezing is no exception. The charts on the following pages show you, at-a-glance, how to prepare, freeze, store and serve various kinds of dishes.

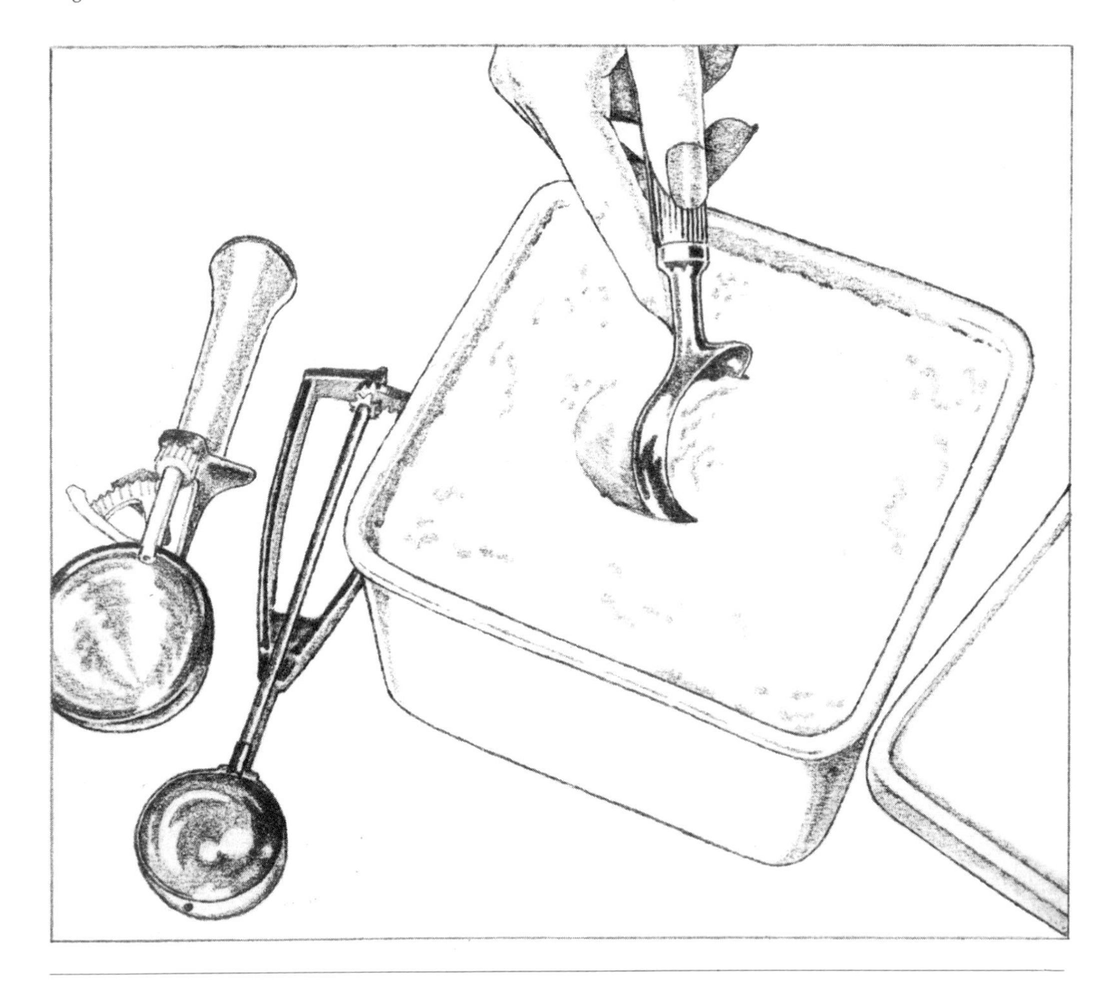

Ice-cream scoops make it easy to serve portions from a large tub.

BASIC KNOW-HOW

Food and storage time	*Preparation*	*Freezing*	*Thawing and serving*
Meat, Raw (leave unstuffed) Beef: 8 months Lamb: 6 months Veal: 6 months Pork: 6 months Freshly minced meat: 3 months Offal: 3 months Cured and smoked meats: 1–2 months Sausages: 3 months	Use good quality, well-hung fresh meat. Removing bones will save space. Butcher in suitable quantities. Place foil or non-stick sheets between individual chops or steaks.	Package carefully in heavy-quality freezer bags. Group in similar types, and overwrap with mutton cloth, stockinette, thin polythene or newspaper, to protect against puncturing and loss of quality.	Most meats may be cooked from frozen (see page 53), but with large joints avoid overcooking meat on outside and leaving it raw at centre. When thawing, use the refrigerator; keeping wrappings on, allow about 6 hours per 450 g (1 lb). Small items such as chops, steaks, can be cooked frozen, but use gentle heat. Partial thawing is necessary before egg-and-crumb coating, etc.
Meat, Cooked Dishes Casseroles, stews, curries, etc.: 2–3 months	Prepare as desired, see that the meat is cooked but not overcooked, to allow for reheating. Do not season too heavily—adjust this at point of serving. Have enough liquid or sauce to immerse solid meat completely. Potato, rice or spaghetti, unless otherwise stated, are best added at point of serving; same applies to garlic and celery.	When mixture is quite cold, transfer to rigid cartons; for dishes with a strong smell or colour, inner-line cartons with polythene bags, use foil dishes or freeze in foil-lined cook-ware. See preformers, page 32.	Reheat food from cartons or polythene bags in a saucepan or casserole dish. Pre-shaped foil-wrapped mixtures can be reheated in the original dish. When reheating in a casserole, allow at least 1 hour for heating through in the oven at 200°C (400°F) mark 6, then if necessary reduce heat to 180°C (350°F) mark 4, for 40 minutes and leave until really hot. Alternatively, heat gently in a pan, simmering until thoroughly heated. Or thaw before reheating.
Meat, Roast Sliced with gravy: 3 months without gravy: 2 months	Joints can be roasted and frozen for serving cold—don't over-cook. Reheated whole joints are not very satisfactory. Sliced and frozen cooked meat tends to be dry when reheated.	Best results are achieved by freezing whole joint, thawing, then slicing prior to serving. But small pieces can be sliced and packed in polythene if required to serve cold, or put in foil containers and covered with gravy, if to be served hot.	Allow plenty of time for thawing-out—about 4 hours per 450 g (1 lb) at room temperature, or double that time in the refrigerator, in the wrapping. Sliced meat requires less time.
Meat Loaves, Pâtés 1 month	Follow regular recipe. Package in the usual way, after cooling rapidly. Keep for minimum time.	When quite cold, remove from tin, wrap and freeze.	Thaw preferably overnight. A change in texture and flavour is sometimes noticeable.

BASIC KNOW-HOW—*continued*

Food and storage time	*Preparation*	*Freezing*	*Thawing and serving*
Poultry and Game Chicken: 12 months Duck: 4–5 months Goose: 4–5 months Turkey: 6 months Giblets: 2–3 months Game birds: 9 months Venison: 12 months	Use fresh birds only: prepare and draw in the usual way. Do not stuff before freezing. Cover protruding bones with non-stick paper or foil. Hang game desired time before freezing.	Pack trussed bird inside polythene bag and exclude as much air as possible before sealing. Freeze giblets separately. If wished, freeze in joints, wrap individually, and then overwrap.	Thaw in wrapping. Thaw a small bird in refrigerator overnight; birds up to 1.8 kg (4 lb) up to 12 hours; 1.8–2.7 kg (4–6 lb) up to 15 hours. Thaw larger birds at room temperature; up to 4.5 kg (10 lb) up to 18 hours; up to 9 kg (20 lb) 30 hours. Joints 6 hours.
Fish, Uncooked Whole (salmon, freshwater fish)	Must be really fresh—within 12 hours of the catch. Wash and remove scales by scraping tail-to-head with back of knife. Gut. Wash thoroughly under running water. Drain and dry on a clean cloth.	For best results, place whole fish unwrapped in freezer until solid. Remove, dip in cold water. This forms thin ice over fish. Return to freezer; repeat process until ice glaze is 0.5-cm ($\frac{1}{4}$-inch) thick. Wrap in freezer wrap; support with a thin board.	Allow to thaw for 24 hours in a cool place before cooking. Once thawed, use promptly.
Fish steaks Oily fish including salmon: 2 months White fish: 3 months Smoked salmon: 2–3 months Caviar: do not freeze	Prepare in the usual way.	Separate steaks with double layer of cling film; overwrap in foil or freezer wrap.	Should be cooked from frozen.
Shellfish 1 month	Advisable only if you can freeze the fish within 12 hours of being caught.		
Fish, Cooked Pies, fish cakes, croquettes, kedgeree, mousse, paellas: 1–2 months	Prepare according to recipe, but be sure fish is absolutely fresh. Hard-boiled eggs should be added to kedgeree before reheating.	Freeze in foil-lined containers, remove when hard, then pack in sealed bags.	Either slow-thaw in refrigerator or put straight into the oven at 180°C (350°F) mark 4, to heat, depending on the type of recipe.
Sauces, Soups, Stocks Sauces, soups: 3 months Stock: 6 months	All are very useful as standbys in the freezer.	When cold, pour into rigid containers, seal well and freeze.	Either thaw for 1–2 hours at room temperature, or heat immediately until boiling point is reached.
Pizza, Unbaked Up to 3 months	Prepare traditional yeast mixture to baking stage. Wrap in foil or polythene.	Freeze flat until solid, then overwrap in ones, twos, threes or fours.	Remove packaging and place frozen in cold oven set at 230°C (450°F) mark 8, and bake for 30–35 minutes.

Pizza, Baked Up to 2 months	Bake traditional yeast mixture in usual way.	Package in foil or polythene and freeze as for unbaked pizza.	Remove packaging and place frozen in a preheated oven at 200°C (400°F) mark 6, for about 20 minutes or leave in packaging at room temperature for 2 hours before reheating as above for 10–15 minutes.
Pastry, Uncooked 3 months	Roll out to size required (or shape into vol-au-vent cases). Freeze pie shells unwrapped until hard, to avoid damage. Use foil plates or take frozen shell out of dish after freezing but before wrapping. Discs of pastry can be stacked with waxed paper between. *Note*: Unshaped shortcrust pastry takes about 3 hours to thaw before it can be rolled out. Bulk flaky and puff—prepare up to the last rolling; pack in polythene bags or extra-thick foil and overwrap. To use, leave for 3–4 hours at room temperature, or overnight in refrigerator.	Stack pastry shapes with two pieces of cling film or freezer wrap between layers, so that if needed, one piece can be removed without thawing the whole batch. Place the stack on a piece of cardboard, wrap and seal.	Thaw flat discs at room temperature, fit into pie plate and proceed with recipe. Unbaked pie shells or flat cases should be returned to their original container before cooking: they can go into oven from the freezer (ovenproof glass should first stand for 10 minutes at room temperature); add about 5 minutes to normal baking time.
Pastry, Cooked 3 months	Prepare as usual. Empty cases freeze satisfactorily, but with some change in texture. Prepare pies as directed (using an aluminium foil dish). Brush pastry cases with egg white before filling. Cool completely before freezing.	Wrap carefully—very fragile. Protect the tops of pies with an inverted paper or aluminium pie plate, then wrap and seal.	Leave pies at room temperature for 2–4 hours, depending on size. If required hot, reheat in the oven. Flan cases should be thawed at room temperature for about 1 hour, refresh if wished.
Pastry Pies, Uncooked Double crust: 3 months	Prepare pastry and filling as required. Make large pies in a foil dish or plate, or line an ordinary dish or plate with foil and use as a preformer. Make small pies in patty tins or foil cases. Do not slit top crust of fruit pies.	Freeze uncovered. When frozen, remove small or preformed pies from containers and pack all pies in foil or polythene bags.	Unwrap unbaked fruit pies and place still frozen in the oven at 220°C (425°F) mark 7, for 40–60 minutes, according to type and size. Slit tops of double crusts when beginning to thaw. (Ovenproof glass should first stand for 10 minutes at room temperature.) Add a little to cooking time.
Top crust: 3 months	Prepare pie in usual way; cut fruit into fairly small pieces and blanch if necessary; toss with sugar; or use cold cooked savoury filling. Cover with pastry. Do not slit crust.	Use ovenproof glass or foil dishes. Wrap in foil or polythene film, protecting as for cooked pies.	Unwrap, place in a preheated oven and bake, allowing extra time. Cut a vent in the pastry when it begins to thaw.

BASIC KNOW-HOW—*continued*

Food and storage time	*Preparation*	*Freezing*	*Thawing and serving*
Pastry Pies, Uncooked (cont'd) Biscuit pie crust: 2 months	Shape in a sandwich tin or pie plate, lined with foil or waxed paper. Add filling if suitable.	Freeze until firm, then remove from tin in the foil wrapping and pack in a rigid container.	Filled: serve cold; thaw at room temperature for 6 hours.
Pancakes, Unfilled 2 months	Add 15 ml (1 tbsp) corn oil to a basic 100 g (4 oz) flour recipe. Make pancakes, and cool quickly on a wire rack. Interleave with lightly oiled non-stick paper or polythene film. Seal in polythene bags or foil.	Freeze quickly.	To thaw: leave in packaging at room temperature for 2–3 hours, or overnight in the refrigerator. For quick thawing, unwrap, spread out separately and leave at room temperature for about 20 minutes. To reheat, place stack of pancakes wrapped in foil in the oven at 190°C (375°F) mark 5, for 20–30 minutes. Alternatively, separate pancakes and place in a lightly greased heated frying pan, allowing $\frac{1}{2}$ minute for each side.
Pancakes, Filled 1–2 months	Only choose fillings suitable for freezing. Don't over-season.	Place filled pancakes in a foil dish, seal and overwrap.	Place frozen, covered, in oven at 200°C (400°F) mark 6, for about 30 minutes.
Sponge Puddings, Uncooked 1 month	Make in the usual way. Use foil or polythene basins, or line ordinary basins with greased foil.	Seal basins tightly with foil, overwrap and freeze at once. To freeze pudding mixture in preformed foil, remove from basins when frozen, then overwrap. *Note*: allow room at this stage for later rising.	Remove packaging, cover top with greased foil and place, frozen, to steam—900-ml ($1\frac{1}{2}$-pint) size takes about $2\frac{1}{2}$ hours. Return preformed pudding mixture to its original basin.
Sponge Puddings, Cooked 3 months	Prepare and cook in the usual way. Cool thoroughly, cover with foil and overwrap.	Freeze quickly.	As above—a 900 ml ($1\frac{1}{2}$ pint) pudding takes about 45 minutes to thaw and reheat.

Ice Cream 3 months Commercially-made: 1 month	Either homemade or bought ice creams and sorbets can be stored in the freezer.	Bought ice creams should be rewrapped in moisture-proof bags before storing. Homemade ones should be frozen in moulds or rigid containers and overwrapped.	Put in refrigerator for short time to soften a little. Some 'soft' bought ice cream can be used from freezer, provided it is not kept in the coldest part. Warn children that ice lollipops and water ices eaten straight from the freezer are dangerous—they may burn your mouth.
Sweets Mousses, creams, etc. 2 months	Make as usual. Any dishes used in freezer must be freezer-proof.	Freeze, unwrapped, in foil-lined container until firm, then remove container, place sweet in polythene bag, seal and return to freezer.	Unwrap preformed mixtures and return to dish, then thaw in refrigerator for about 6 hours, or at room temperature for about 2 hours.
Cream Fresh: 3 months Commercially-frozen: up to 1 year	Use only pasteurised, with a 35 per cent butter-fat content or more (ie double or whipping cream). Whipped cream may be piped into rosettes on waxed paper. Best results are achieved with half-whipped cream with a little sugar added—5 ml (1 level tsp) to 150 ml ($\frac{1}{4}$ pint).	Transfer cream to suitable container, eg waxed carton, leaving space for expansion. Freeze rosettes unwrapped; when firm, pack in a single layer in foil.	Thaw in refrigerator, allowing 24 hours, or 12 hours at room temperature. Put rosettes in position as decoration before thawing, or they cannot be handled; these take less time to thaw.
Cakes, Cooked Including sponge flans, swiss rolls and layer cakes: 3 months (frosted cakes lose quality after 2 months; since aging improves fruit cakes, they may be kept longer)	Bake in usual way. Leave until cold on a wire rack. Swiss rolls are best rolled up in cornflour, not sugar, if to be frozen without a filling. Do not spread or layer with jam before freezing. Keep flavourings to a minimum and go lightly with spices.	Wrap plain cake layers separately, or together with cling film or waxed paper between layers. Freeze frosted cakes (whole or cut) unwrapped until frosting has set, then wrap, seal and pack in boxes to protect icing.	Frosted cakes: unwrap before thawing, then the wrapping will not stick to the frosting when thawing. Cream cakes: may be sliced while frozen, for a better shape and quick thawing. Plain cakes: leave in package and thaw at room temperature. Un-iced layer cakes and small cakes thaw in about 1–2 hours at room temperature: frosted layer cakes take up to 4 hours.
Cake Mixtures, Uncooked 2 months	Whisked sponge mixtures do not freeze well uncooked. Put rich creamed mixtures into containers, or line the tin to be used later with greased foil, add cake mixture and freeze uncovered. When frozen, remove from tin, package in foil and overwrap.	Return to freezer.	To thaw, leave at room temperature for 2–3 hours, then fill tins to bake. Preformed cake mixtures can be returned to the original tin, without wrapping but still in foil lining. Place frozen in pre-heated oven and bake in usual way, but allow longer cooking time.

BASIC KNOW-HOW—*continued*

Food and storage time	*Preparation*	*Freezing*	*Thawing and serving*
Scones and Teabreads 3 months	Bake in usual way.	Freeze in polythene bag in convenient units for serving.	Thaw teabreads in wrapping at room temperature for 2–3 hours. Tea scones: cook from frozen, wrapped in foil, at 200°C (400°F) mark 6, for 10 minutes. Girdle scones: thaw 1 hour. Drop scones: thaw 30 minutes or cover and bake for 10 minutes.
Croissants and Danish Pastries Unbaked, in bulk: 6 weeks Baked: 4 weeks	Unbaked: Prepare to the stage when all the fat has been absorbed, but don't give the final rolling. Baked: Bake as usual.	Unbaked: Wrap in airtight polythene bags and freeze at once. Baked: Pack in polythene bags.	Leave in polythene bag, but unseal and re-tie loosely, allowing space for dough to rise. Preferably thaw overnight in a refrigerator, or leave for 5 hours at room temperature. Complete the final rolling and shaping, and bake. Baked: Loosen wrapping and thaw as for unbaked.
Biscuits, Baked and Unbaked 3 months	Prepare in the usual way. Rich mixtures—ie with more than 100 g (4 oz) fat to 450 g (1 lb) flour—are the most satisfactory.	Either baked or unbaked, pack carefully. Wrap rolls of uncooked dough, or pipe soft mixtures into shapes, freeze and pack when firm. Allow cooked biscuits to cool before packing.	Thaw uncooked rolls of dough slightly; slice off required number of biscuits and bake. Shaped biscuits can be cooked direct from frozen: allow 7–10 minutes extra cooking time. Cooked biscuits may require crisping in warm oven.
Bread 2 weeks	Freshly-baked bread, both bought and homemade, can be frozen. Crisp, crusty bread stores well up to 1 week, then the crust begins to 'shell off'.	Bought bread may be frozen in original wrapper for up to 1 week; for longer periods, seal in foil or polythene. Homemade bread: freeze in foil or polythene bags.	Leave to thaw in the sealed polythene bag or wrapper at room temperature 3–6 hours, or overnight in the refrigerator, or leave foil-wrapped and crisp in oven at 200°C (400°F) mark 6, for about 45 minutes. Sliced bought bread can be toasted from frozen.
Bought Part-Baked Bread and Rolls 4 months	Freeze immediately after purchase.	Leave loaf in the bag. Pack rolls in heavy-duty polythene bags and seal.	To use, place frozen unwrapped loaf in oven at 220°C (425°F) mark 7, for about 40 minutes. Cool for 1–2 hours before cutting. Rolls: place frozen unwrapped in oven at 200°C (400°F) mark 6, for 15 minutes.

Sandwiches 1–2 months	Most types may be frozen, but those filled with hardboiled eggs, tomatoes, cucumber or bananas tend to go tasteless and soggy.	Wrap in foil, then in polythene bag.	Thaw unwrapped at room temperature or in the refrigerator. Times vary according to size of pack. Cut pinwheels, sandwich loaves, etc., in portions when half-thawed.
Marmalade Oranges 6 months *Note*: Useful if it's not convenient to make marmalade when Seville oranges are in season.	Wash, dry and freeze Seville oranges whole, or prepare marmalade to cooked pulp stage—ie before addition of sugar.	Pack whole oranges in polythene bags; pulp in suitable containers.	It is not recommended to thaw whole frozen fruit in order to cut it up before cooking as some discoloration often occurs—use whole fruit method for marmalade. It is advisable to add one-eighth extra weight of Seville or bitter oranges or tangerines when freezing for subsequent marmalade making in order to offset pectin loss.
Fats Butter, salted: 3 months Butter, unsalted: 6 months Margarine: as butter Fresh shredded suet: 6 months	Always buy fresh stock (farmhouse butter must be made from pasteurised cream).	Overwrap in foil in 225–450 g ($\frac{1}{2}$–1 lb) quantities.	Allow to thaw in refrigerator. 4 hours for a 225-g ($\frac{1}{2}$-lb) block.
Commercially Frozen Foods Up to 3 months as a rule *Note*: The times quoted by the manufacturers are often less than those given for home-frozen foods, because of the handling in distribution, before the foods can reach your own freezer.	No further preparation, etc., needed, except for ice cream, which should be overwrapped if it is to be kept for longer than 3 weeks.	Follow directions on packet.	
Herbs Up to 6 months	Wash and trim if necessary. Dry thoroughly.	Freeze in small bunches in a rigid foil container or polythene bag. Alternatively, herbs, especially parsley, can be chopped before freezing.	Can be used immediately. Crumble while still frozen.
Eggs separated 8–10 months	Freeze only fresh eggs—yolks and whites separately.	Pack in waxed or rigid containers. Yolks—to every 6 yolks add 5 ml (1 level tsp) salt or 10 ml (2 level tsps) sugar; to single yolks add 2.5 ml ($\frac{1}{2}$ level tsp) salt or sugar.	Thaw in refrigerator or rapidly thaw at room temperature for about $1\frac{1}{2}$ hours.

BASIC KNOW-HOW—*continued*

Food and storage time	*Preparation*	*Freezing*	*Thawing and serving*
Milk 1 month	Ordinary pasteurised milk does not freeze well. Homogenised is satisfactory.	Pack in rigid containers; allow 2.5-cm (1-inch) headspace. *Do not freeze in bottle.*	Thaw in refrigerator. Thawing may be accelerated if milk is to be used in cooking.
Yoghurt 6 weeks	Fruit yoghurts are satisfactory but natural yoghurts do not freeze well. Some can be bought ready-frozen.	Freeze in retail cartons.	Thaw for about 1 hour at room temperature.
Cheese 3–6 months	Soft cheeses and cream cheeses are suitable for freezing. Hard cheeses become crumbly if stored for too long but are fine grated for cooking. Cottage cheese not suitable for freezing.	Wrap in freezer film or polythene bag.	Thaw for 24 hours in refrigerator and allow to come to room temperature before serving. Use grated cheese straight from frozen.

Keep your freezer stocked with interesting impromptu suppers, such as Chilli Con Carne, Crêpes Ratatouille and Wholewheat Pizza.

FREEZER STANDBYS

The trouble with the last-minute trimmings is that there's so much else to do in that last minute. You can save yourself time and trouble by having the trimmings ready prepared, as well as other less glamorous standbys like breadcrumbs and seasoned flour, baker's yeast and well-flavoured stock.

Maître d'Hôtel and Garlic Butters

450 g (1 lb) unsalted butter
grated rind of 2 lemons
60 ml (4 tbsp) lemon juice
120 ml (8 level tbsp) chopped fresh parsley
salt and freshly ground pepper
4 garlic cloves, skinned and crushed

Ideal for topping steaks, chops and other grills such as fish.

1 Cream unsalted butter until smooth. Beat in the grated lemon rind, lemon juice, parsley and a little seasoning.

2 Divide into two, and add the garlic to one portion. Chill both portions slightly in the freezer or refrigerator until of a manageable consistency.

3 Form each piece into a long sausage shape between sheets of waxed or non-stick paper. Re-firm in the fridge or freezer, and cut into 50 g (2 oz) portions. Wrap each in foil. Alternatively, freeze the butter roll whole and cut off slices as required. Make a neat parcel, seal and label. Pack the firm labelled parcels together in one polythene bag. Use straight from the freezer.

Mint Butter

40 g ($1\frac{1}{2}$ oz) fresh mint
75 ml (5 level tbsp) fresh parsley
100 g (4 oz) unsalted butter
lemon juice
pinch of salt

1 Wash and drain the mint and parsley. Put in a pan with 60 ml (4 tbsp) water; cook uncovered for 5 minutes, by which time most of the water will have evaporated. Rub the mixture through a sieve.

2 Cream butter and work in the mint purée, a good squeeze of lemon juice and a pinch of salt. Form into a long sausage shape between sheets of waxed or non-stick paper. Re-firm in the fridge or freezer, and cut into 50 g (2 oz) portions. Wrap each in foil. Alternatively, freeze the butter roll whole and cut off slices as required. Make a neat parcel, seal and label. Pack the firm labelled parcels together in one polythene bag. Use straight from the freezer.

Butter Balls

These take time to make when needed, but a few stored in the freezer—even those left over from a party—are handy for almost instant use.

Heavenly puddings straight from the freezer—spectacular Almond and Coffee Meringue Cake, Butterscotch Ice Cream with toasted almonds and luscious Chocolate Supreme.

Many useful cookery standbys keep well in the freezer—all packed in suitably-sized containers and labelled. They can include breadcrumbs, egg yolks and whites (separated), lemon and orange peel (either grated or in slivers), fresh herbs and white sauce. See the suggestions we list in this chapter for other useful, time-saving ideas.

Stock

Particularly if you use a pressure cooker, the making of good-quality homemade stock is a simple affair. Even chicken bones, with skin, giblets, etc, are well worth converting into a jellied stock; cooked meat bones, browned carrot, onion and celery, bacon rinds and mushroom peelings make a very useful concentrated stock. Or you can start from scratch with raw beef and veal bones. When the stock is ready, pour some of it into ice-cube trays—useful when only a small amount is needed—or use preformers. Freeze.

Incidentally, the freezer is the ideal place to store a few odd trimmings and bones left when you have prepared or carved a bird, to wait till you have enough accumulated, or enough time, to turn them into stock.

Parsley

Parsley and other fresh herbs may be frozen unblanched. Freeze clean parsley heads (pack the stalks separately—these are often required as a flavouring in recipes) and crumble while still frozen—to 'chop' the parsley. Alternatively, chop large amounts of crisp fresh parsley, or other herbs, divide between small containers (for example, ice-cube trays) and freeze until firm. Leave in the small containers or take from the ice-cube trays and pack carefully in rigid containers. Use straight from the freezer as an addition to stews, casseroles, sauces, etc.

Sauce Cubes

So convenient when only a small amount of sauce is needed at any one time. Make up a well-flavoured curry, tomato, espagnole or other sauce, using for preference a homemade stock base rather than instant cubes. Cool, and pour into ice-cube trays. Freeze unwrapped, divide into cubes and freeze in polythene bags. To use, select the number of cubes required and reheat slowly with a little extra liquid if necessary.

Garlic and Other Savoury Breads

French or Vienna bread loaves
flavoured butter

These go down so well when served with soups on a buffet occasion that it's worth stocking up before a party.

1 Make 2.5-cm (1-inch) cuts along French or Vienna loaves to within 1 cm ($\frac{1}{2}$ inch) of the base, open up and spread generously with creamed butter flavoured with garlic, cheese or herbs.

2 Wrap the loaves tightly in foil, then freeze.

3 To thaw and 'refresh', place the frozen, foil-wrapped loaf in the oven at 200°C (400°F) mark 6; a French stick takes about 30 minutes, and a Vienna loaf about 40 minutes.

Horseradish Cream

150 ml ($\frac{1}{4}$ pint) double cream
15 ml (1 tbsp) creamed horseradish
lemon juice

1 Whip the cream and fold in the horseradish. Add a good squeeze of lemon juice and turn the mixture into a small serving dish or rigid foil container.

2 Cover with extra thick or two layers of foil and overwrap with a polythene bag. Seal and label.

3 To use, unwrap, but leave the top loosely covered, and allow to thaw for 6 hours in the refrigerator. Dust with paprika pepper before serving.

Mint Sauce

about 225 g (8 oz) fresh mint
a little vinegar
25–50 g (1–2 oz) caster sugar

The leaves may be frozen whole, or chopped as for other herbs, and used to make up the sauce at the point of serving. However, you can make an almost instant sauce.

1 Chop enough mint to give 225 g (8 oz), moisten with a little vinegar and add caster sugar.

2 Freeze in small amounts—ice-cube trays make ideal containers. When the mixture is frozen, pack the cubes in polythene bags or wrap individually in foil and then mass together in an overwrap; seal and label.

3 To use, add more vinegar, or one-third water and two-thirds vinegar, and leave to thaw.

4 For a brighter colour, moisten before freezing with boiling water instead of cold vinegar, and proceed as above, adding undiluted vinegar to thaw.

It saves space if you freeze stock rather than bones and chopped parsley rather than sprigs. It means they are also quicker to use.

Cranberry Sauce

175 g (6 oz) sugar
225 g (8 oz) fresh cranberries, or frozen, thawed

1 Simmer sugar and 150 ml ($\frac{1}{4}$ pint) water in a pan until the sugar dissolves.
2 Add cranberries and cook over a medium heat for about 10 minutes, until the fruit has broken up; cool.
3 Pot leaving a small headspace, cover with foil and freeze.

Cranberry Relish

225 g (8 oz) fresh cranberries
1 apple, cored and finely chopped, not peeled
1 orange
1 lemon
225 g (8 oz) sugar

Traditionally this belongs to the turkey, but it also complements pork and gammon.
1 Into the goblet of a powerful blender put the fresh cranberries and the chopped apple.
2 Remove the pips from the orange and lemon; mince the flesh finely and stir into the cranberry mixture, together with the sugar.
3 Pot, leaving a small headspace, cover with foil and freeze.

Breadcrumbs

To prepare fresh breadcrumbs in quantity, especially with the aid of a blender, is a real time-saver. When frozen the crumbs remain separate, and the required amounts can be easily removed. Pack in polythene bags, sealed containers, or even screw-topped jars.

There is no need to thaw crumbs to be used for stuffings, pudding or sauces, but for coating fried foods leave them at room temperature for 30 minutes.

Delicious herbs such as mint, chives, parsley, sage and rosemary can all be frozen on the stalk or chopped and frozen in ice-cube trays. Concentrated mint sauce will also freeze well.

Buttered Crumbs

25 g (1 oz) butter
100 g (4 oz) fresh breadcrumbs

A delicious crisp savoury topping for sauce-coated au gratin dishes, or for sweet dishes if sugar is added.

1 Melt butter in a frying pan, stir in breadcrumbs, until blended, then fry slowly until evenly browned and golden.

2 When cold, pack in polythene bags; freeze.

Cheese Titbits

100 g (4 oz) plain flour
a pinch of salt
50 g (2 oz) butter or block margarine
50 g (2 oz) Cheddar cheese, finely grated
a little beaten egg or water to mix
milk to glaze
poppy seeds to decorate (optional)

When making cheese pastry for savouries or flans, roll out any leftover trimmings and cut into fancy shapes to make these titbits, which are useful for decorating soups or to serve separately; if topped with a little pâté they can accompany an apéritif. Or you can make the pastry specially for this purpose.

1 Mix the flour and salt together and rub in the fat as for shortcrust pastry, until the mixture resembles fine breadcrumbs. Mix in the cheese and add the egg or water, stirring until the ingredients begin to stick together.

2 Then with one hand collect the dough together and knead very lightly to give a smooth dough.

3 Roll out as for shortcrust pastry and cut into fancy shapes. Brush the tops with a little milk and if desired sprinkle with poppy seeds.

4 Bake at 200°C (400°F) mark 6 until golden brown.

5 Pack and freeze as usual.

Fried Bread Croûtons

These and fried bread shapes for canapés and party snacks freeze well and can be prepared in quantity for up to 1 month's supply. When frozen, they remain separate for easy handling.
1 Cut the bread slices into 0.5–1-cm ($\frac{1}{4}$–$\frac{1}{2}$-inch) cubes, or into fancy shapes, eg, stars.
2 Fry in shallow or deep fat until crisp and golden; drain well. Pack as for breadcrumbs and freeze.
3 To thaw and 'refresh', place uncovered but still frozen, in the oven at 200°C (400°F) mark 6 for 5 minutes.

Duxelles

This paste or mince, made of mushrooms, or mushroom peelings and stalks, is used for flavouring sauces, soups, ragoûts, stuffings, stews and so forth. For the recipe see page 79.

Fruit Ice Cubes

Fresh fruit-flavoured cubes are invaluable at party times for chilling a drink or for adding, say, the orange, lemon or grapefruit flavour without unduly diluting a drink. Try also tomato cubes (made by chilling a tomato cocktail mixture), or experiment with fresh orange juice cubes added to tomato juice. To prevent cubes sticking together during storage, when you pack them in a polythene bag, place them in the bag, pour over them a little soda water and shake bag well, to coat each cube; seal and label.

To make decorated ice cubes, add the juice of 1 lemon to 300 ml ($\frac{1}{2}$ pint) water and pour into ice-cube trays. Slice the remaining lemon, then cut the slices into quarters and pop a small piece into each section. Freeze in the usual way, turn out and pack. The same treatment can be given to oranges or limes.

Another pleasant trick is to pop a Maraschino cherry into each section of the ice-cube tray, then fill up with soda water—good for punches and apéritifs.

A little soda water in the polythene bag, shaken well, will prevent ice cubes from sticking together in the freezer.

Whether sweet or savoury, a ready-made selection of trimmings and garnishes will take the panic out of last-minute finishing touches.

Citrus Fruit Rind

When only the juice of the fruit or just the flesh is needed for a recipe, the rind is often wasted. It's a good idea to thinly pare the rind free of pith, or grate the rind before squeezing the juice out or using the flesh (it's easier to do it first rather than afterwards). Pack the rind by itself, closely wrapped in foil or packed in small containers, or in a jar with sugar, to give lemon or orange sugar—a natural flavouring for cakes, biscuits, breads and so on. The strips of rind can be popped—still frozen—into cold drinks.

Lemon Curd

Freezing will extend the life of your favourite lemon curd for up to 3 months.

Uncooked Jam

550 g ($1\frac{1}{4}$ lb) raspberries or strawberries, hulled
900 g (2 lb) caster sugar
$\frac{1}{2}$ a bottle of liquid pectin, such as Certo
30 ml (2 tbsp) lemon juice

In preserves made without cooking, the fresh fruit flavour is superb and the set is more like that of a conserve, not firm like boiled jam.

1 Crush raspberries or strawberries with caster sugar. Leave to stand for about 1 hour in a warm kitchen, stirring occasionally. All the sugar should be dissolved.

2 Now add pectin and lemon juice. Stir for a further 2 minutes.

3 Pour into small dry jars, leaving 1-cm ($\frac{1}{2}$-inch) headspace, and cover with foil. Leave in the warm kitchen for a further 48 hours, then freeze.

MAKES ABOUT 1.6 Kg ($3\frac{1}{2}$ lb)

Meringue Shells

2 egg whites
50 g (2 oz) sugar
50 g (2 oz) caster sugar

Can be used straight from the freezer for decorating vacherins, etc.

1 Line a baking sheet with a sheet of non-stick paper.
2 Whisk egg whites until very stiff, add sugar and whisk again until the mixture regains its former stiffness. Lastly fold in caster sugar very lightly, using a metal spoon.
3 Put into a forcing bag fitted with a large fluted star nozzle. Pipe 'shell' shapes on to the lined baking sheet.
4 Bake in a very cool oven, at 130°C (250°F) mark $\frac{1}{4}$, until the meringues are firm and crisp, but still white. (If they begin to brown, prop the oven door open a little.)
5 Cool before packaging in rigid plastic containers for freezing.

Mocha Sauce

200 g (8 oz) plain dark chocolate
50 g (2 oz) butter
45 ml (3 level tbsp) instant coffee
90 ml (6 tbsp) milk

To serve with ice cream, profiteroles, chilled pears, bananas, filled meringues, etc.

1 Melt chocolate; add butter and coffee blended with the milk.
2 Heat in a double pan, or in a bowl over hot water, until smooth —stir constantly.
3 Pour into a 400-ml ($\frac{3}{4}$-pint) rigid container. When cold, cover, overwrap, seal and label, then freeze.
4 To use, reheat in a double pan or over hot water.

MAKES ABOUT 300 ml ($\frac{1}{2}$ pint)

Chocolate Shapes and Caraque

These very delicate decorations will freeze quite well, so it's worth keeping a few in a rigid box in the freezer for special occasions.

Squares: Melt some plain chocolate. Using a palette knife, spread it in a thin, even layer over a sheet of non-stick paper, but don't take it quite to the edges. When firm, but not brittle, cut squares of the required size, using a sharp knife, with a ruler as guide. When it is firm, place the paper over the edge of a table, pull it down, and ease the squares off one at a time.

Leaves: Melt the chocolate. Collect some small rose leaves, preferably with the stem attached. Spread a little of the chocolate over the *under*side of each leaf, using a teaspoon. Place on non-stick paper and leave to set. When hard, pull the leaves away, starting from the stem. Work gently but quickly, to avoid the chocolate melting.

Caraque: Melt the chocolate in the usual way and spread it thinly on a cold work surface. When it is just on the point of setting, shave it off in curls with a thin, sharp knife. Some of the softer blocks of chocolate or special chocolate covering can be 'curled' by using a potato peeler on the flat side of the block.

FREEZER RECIPES WHICH WORK

The freezer four-day cook-in

Set aside a quiet, uninterrupted day for each cook-in. This means planning ahead to clear the decks of children and household pets. At the end of the day you may then glow with a sense of achievement—and from a glass of what you fancy. You'll deserve it!

We've allowed six hours per session, starting from scratch but with all ingredients to hand: the shopping done in advance. Average oven space has been utilised, with timing arranged to avoid over-loading. Clear the equivalent of one deep shelf-space per session in the freezer to take cooked dishes and those for open-freezing. All recipes mentioned in the cooking plans are given after.

Practical points: Never put into the freezer food that's even lukewarm. Pre-chill in fridge when necessary. Wrap all food correctly to exclude as much air as possible. Allow headspace for liquids since they expand during freezing. Portion according to family needs: one or two portions are often more useful than bigger amounts. Remember it's easy to purée unseasoned mini-portions for babies. Label everything, and include thawing and reheating directions for later convenience.

Shopping List

Butcher **225 g (8 oz) pork sausagemeat; 225 g (8 oz) streaky bacon rashers; 225 g (8 oz) cooked ham.**
Greengrocer **450 g (1 lb) onions; 100 g (4 oz) button mushrooms; 1 orange.**
Grocer **Two 1.3-kg (3-lb) bags plain flour; 450 g (1 lb) each of block margarine and lard; 275 g (10 oz) butter; fat for deep frying; 225 g (8 oz) Cheddar cheese; 396-g (13-oz) can apricot halves; 226-g (8-oz) can tomatoes; large tube tomato purée; 8 size 4 eggs; 3 size 2 eggs.**
Dairy **568 ml (1 pint) milk; 142 ml ($\frac{1}{4}$ pint) double cream.**
Store cupboard **Dry sherry; red wine; dried mixed herbs; caster sugar; nutmeg.**

PASTRY DAY

Shortcrust, the most popular of all pastry, freezes excellently; it's a most useful item for the freezer. Delicious flans can be filled and frozen cooked or uncooked. Empty pastry cases are wonderful standbys for filling later with what-you-will; they're rather fragile, so treat with respect when packing. For interest and variety we've included a batch of choux pastry.

All pastry dishes from the freezer, even if they're to be served cold, benefit from being refreshed in the oven before being left to cool again, provided the other ingredients won't spoil.

Cooking plan

1 In the morning, switch the freezer to fast-freeze; then prepare two batches of shortcrust pastry. (If large enough bowls aren't available, do this in four batches.) Weigh off into bowls four separate 450 g (1 lb) lots of dry mix and bag up the remainder for limited fridge or long-term freezer storage.

2 Make up each 450-g (1-lb) dry mix into shortcrust pastry. Put one portion in a polythene bag. Divide one portion in half and roll out to line two 23-cm (9-inch) foil flan cases. Repeat, using a second portion to give four flan cases in all. Keep trimmings.

3 Ready for making up, skin and finely chop 450 g (1 lb) onions, grate 225 g (8 oz) cheese; trim fat from ham, chop finely.
4 Make up the Sausage Jalousie (pasty), using the third portion of pastry, bake and cool.
5 Use the two prepared pastry cases to make the Bacon Quiche and Sweet Egg Custard Tart. Cool quiche and open-freeze tart.
6 Roll out the remaining (bagged) pastry dough, including trimmings and use to line about eight 70-cm (4-inch) individual loose-bottomed fluted flan tins. Open-freeze: these can be removed from tins after freezing and stored in the freezer if carefully packed, then returned to tins for baking.
7 In the afternoon, make up two basic amounts of choux pastry in two separate pans, preferably using a hand mixer.
8 Make about twelve Orange Aigrettes using half the mixture from one pan. Prepare Savoury Aigrettes with the remaining choux.
9 Using the second batch of choux pastry, prepare savoury gougère (flan), the choux ring, or buns.
10 Make the savoury filling and apricot sauce for plain Aigrettes. Pipe whirls of lightly whipped cream on to a non-stick lined baking sheet and open-freeze.
11 Pack and freeze all cold pastry items (including large empty flan shells): filling and sauce in small containers.

Basic Shortcrust Pastry

900 g (2 lb) plain flour
5 ml (1 level tsp) salt
225 g (8 oz) block margarine
225 g (8 oz) lard

1 Place the flour and salt in a bowl and rub in the fat until the mixture resembles breadcrumbs.
2 Weigh off separate 450 g (1 lb) lots of dry mix into bowls. (Bag up any remainder for limited fridge or long-term freezer storage.)
3 Make each 450 g (1 lb) dry mix into pastry, using about 45 ml (3 tbsp) cold water. Knead each piece of pastry on a floured work surface.

MAKES 1.4 Kg (3 lb)

Bacon Quiche

225 g (8 oz) shortcrust pastry, used to line a 23-cm (9-inch) foil flan dish (see above)
100 g (4 oz) streaky bacon, chopped
100 g (4 oz) finely chopped onion
1 egg, size 2
142 ml ($\frac{1}{4}$ pint) milk
2.5 ml ($\frac{1}{2}$ level tsp) dried mixed herbs
salt and freshly ground pepper

1 Place the pastry-lined foil flan dish on a baking sheet.
2 Fry the bacon in its own fat, and when fat starts to run add onion and cook gently until tender. Place in the flan case.
3 Beat the egg and milk together in a bowl with the herbs and seasoning, and pour into the flan case.
4 Roll out the pastry trimmings and use to form a lattice on top of the flan.
5 Bake in the oven at 200°C (400°F) mark 6 for 20–25 minutes until just set. Cool. Pack and freeze.
6 To use, unwrap and reheat from frozen in the foil dish placed on a baking sheet in the oven, at 190°C (375°F) mark 5 for about 30 minutes. If necessary, cover with foil during cooking to prevent overbrowning.

MAKES 3 PORTIONS

Sausage and Bacon Jalousie

225 g (8 oz) pork sausagemeat
100 g (4 oz) streaky bacon, chopped
100 g (4 oz) finely chopped onion
450 g (1 lb) shortcrust pastry (see above)
75 g (3 oz) Cheddar cheese, grated
30 ml (2 level tbsp) tomato purée
a little milk

1 Form sausagemeat into twelve small balls. Fry gently in a frying pan with bacon and onion until golden brown. Cool.
2 Divide pastry in half and roll one piece into rectangle 28 × 18 cm (11 × 7 inches). Roll remaining pastry into a slightly larger rectangle. Cut each in half lengthwise.
3 Mix the sausage balls, bacon and onion with cheese and tomato purée. Spread over the two smaller rectangles placed on a greased baking sheet. Dampen edges.
4 Lay remaining pastry lengths over the filling and neaten edges to give a domed appearance. Knock up edges and pinch together with fingers.
5 With a sharp knife, slash the pastry top crosswise to give slits about 1 cm ($\frac{1}{2}$ inch) apart. Brush with milk.
6 Bake in the oven at 200°C (400°F) mark 6 for about 30 minutes. Cool on a wire rack. Pack separately in foil and freeze.
7 To use, unwrap and reheat from frozen in foil, placed on a baking sheet in the oven at 190°C (375°F) mark 5 for about 45 minutes.

MAKES 2 JALOUSIES FOR SLICING

Sweet Egg Custard Tart

225 g (8 oz) shortcrust pastry, used to line a 23-cm (9-inch) foil flan dish (see above)
2 eggs, size 2
30 ml (2 level tbsp) caster sugar
300 ml ($\frac{1}{2}$ pint) milk
grated nutmeg

1 Place the pastry-lined foil flan case on a baking sheet.
2 Whisk the eggs lightly with the sugar; warm the milk until almost boiling and add to the eggs. Cool.
3 Strain the custard into the pastry case, sprinkle nutmeg over and open freeze until firm. Pack and freeze.
4 To use, unwrap and cook from frozen in the foil dish placed on a baking sheet in the oven at 190°C (375°F) mark 5 for about 1 hour.

MAKES ENOUGH FOR 4–6

Basic Choux Pastry

100 g (4 oz) butter
150 g (5 oz) plain flour
4 eggs, beaten

1 Melt the butter in a pan with 300 ml ($\frac{1}{2}$ pint) water and when just boiling, remove from heat. Add flour all at once and beat gently, just enough to incorporate flour.
2 Using a hand electric mixer, gradually beat in eggs until all are incorporated.

Savoury Aigrettes

$\frac{1}{2}$ batch choux pastry (see above)
50 g (2 oz) ham, chopped
50 g (2 oz) Cheddar cheese, grated
salt and freshly ground pepper

1 Mix the choux pastry, ham and cheese together. Season.
2 Deep fry teaspoonfuls of the mixture, a few at a time, in hot vegetable oil at about 190°C (375°F) until crisp and golden. Drain and cool, then pack and freeze.
3 To use, unwrap, place on baking sheets and refresh in the oven at 200°C (400°F) mark 6 for about 10 minutes.

MAKES ABOUT 12

Sweet Orange Aigrettes

$\frac{1}{2}$ batch choux pastry (see above)
grated rind of $\frac{1}{2}$ an orange

1 Mix choux pastry and grated orange rind together.
2 Deep fry teaspoonfuls of the mixture, a few at a time, in hot vegetable oil at about 190°C (375°F) until crisp and golden. Drain and cool, then pack and freeze.
3 To use, unwrap, place on baking sheet and refresh in oven at 200°C (400°F) mark 6 for about 10 minutes. Dredge with sugar and serve with apricot sauce.

MAKES ABOUT 12

Apricot Sauce

396-g (13-oz) can apricots, with 150 ml ($\frac{1}{4}$ pint) juice
grated rind of $\frac{1}{2}$ an orange
juice of 1 orange
45 ml (3 tbsp) dry sherry
30 ml (2 level tbsp) caster sugar

1 Purée all the ingredients together in a blender. Pack in a rigid container and freeze.
2 To use, bring to the boil slowly in a saucepan.

MAKES 150 fl oz ($\frac{1}{4}$ pint)

Ham Gougère and Choux Ring

1 batch choux paste (see above)
75 g (3 oz) finely chopped onion
100 g (4 oz) mushrooms, wiped and sliced
50 g (2 oz) butter
15 ml (1 level tbsp) plain flour
226-g (8-oz) can tomatoes, with juice
30 ml (2 level tbsp) tomato purée
15 ml (1 tbsp) red wine
175 g (6 oz) ham, chopped
salt and freshly ground pepper
50 g (2 oz) Cheddar cheese, grated
142 ml ($\frac{1}{4}$ pint) double cream

1 Place choux paste in a forcing bag fitted with a 1-cm ($\frac{1}{2}$-inch) plain nozzle and pipe about one-third on to a dampened baking sheet in the shape of a ring about 20.5-cm (8-inch) in diameter or into sixteen buns.
2 Bake in the oven at 220°C (425°F) mark 7 for 15–20 minutes. Cool on a wire rack.
3 Make another ring the same size, by piping another third of the paste on to a non-stick paper-lined baking sheet. Use the remaining paste to fill in the base, then open freeze until firm. Peel off paper, pack and return to freezer. Pack baked ring on a foil plate protected by a second plate and sealed, then freeze.
4 To make the filling, sauté the onion and mushrooms in butter for 3 minutes. Stir in the flour and cook for 2 minutes, then add all the remaining ingredients except the cheese. Pack sauce and cheese separately, then freeze.
5 To use gougère, reheat filling in a pan, spoon into frozen choux base placed on a baking sheet. Top filling with cheese and bake in the oven at 190°C (375°F) mark 5 for about 45 minutes, until puffed up and bubbling.

To use empty baked ring or buns, refresh in oven as for aigrettes. Cool on a wire rack, split and fill with frozen whirls of cream and chopped fruit, dredge with icing sugar. Leave for cream to thaw.

EACH MAKES ENOUGH FOR 4–6

SOUP AND VEGETABLE DAY

Shopping List
Butcher **225 g (8 oz) boiled bacon—hock or collar; 2.3 kg (5 lb) veal bones (optional); 1 chicken (optional).**
Greengrocer **900 g (2 lb) onions; 3 kg (7 lb) old potatoes; 225 g (8 oz) each of parsnips (or celery) and carrots; 900 g (2 lb) tomatoes; 1.1 kg ($2\frac{1}{2}$ lb) button mushrooms.**
Grocer **50 g (2 oz) pearl barley; 225 g (8 oz) butter; 3 eggs.**
Dairy **568 ml (1 pint) milk.**
Baker **1 large uncut loaf.**
Store cupboard **Oil; dry white wine; self-raising flour; shredded suet; nutmeg; dried thyme and sage; chicken stock cubes (or homemade stock); garlic.**

Homemade soups in the freezer are great comforters and money-savers. Also pack away plenty of concentrated stock made from bones; with a pressure cooker this takes little time and is of inestimable value. List all preparation jobs that can be done together. Peeling and chopping vegetables in one go takes far less time than dealing with a couple of ounces for each recipe.

Note Homemade veal stock gives a good flavour to the Vegetable Broth and Potato Soup. You'll need about 2.8 litres (5 pints) from 2.3 kg (5 lb) chopped veal bones. Make this (do not season it) the day before and also 1 litre ($1\frac{3}{4}$ pints) chicken stock.

Cooking plan

1 In morning, skin onions; finely chop all but 175 g (6 oz).
2 Peel and cut up potatoes and keep under cold water.
3 Pare carrots and peel parsnips, keep under cold water.
4 Make soups starting with Vegetable and Bacon Broth. (Chop parsley and freeze in small packs ready for garnishes.) When soups are cool, chill in fridge.
5 In afternoon, remove crusts from loaf, crumb 175 g (6 oz) and dice remainder.
6 Prepare duchesse and croquette potatoes and, when cool, open-freeze.
7 Prepare and cool mushrooms à la grecque and duxelles.
8 Fry the diced bread cubes a few at a time in deep or shallow fat until golden. Drain on kitchen paper.
9 Prepare dumpling dough.
10 Pack soups in amounts likely to be needed, ideally not over 600 ml (1 pint). Freeze.
11 Pack and freeze duchesse and croquette potatoes; mushrooms à la grecque, duxelles and fried croûtons.

Vegetable and Bacon Broth

30 ml (2 tbsp) oil
225 g (8 oz) parsnips or celery, peeled and diced
225 g (8 oz) carrots, peeled and diced
100 g (4 oz) finely chopped onion
225 g (8 oz) boiled bacon, hock or collar, finely diced
2 tomatoes, skinned and chopped
50 g (2 oz) pearl barley
1.1 litres (2 pints) veal bone stock
salt and freshly ground pepper
100 g (4 oz) frozen peas and frozen dumplings to serve (optional)

1 Heat the oil in a large pan and sauté the parsnips or celery, carrots and the onion quickly, without browning.
2 Add the bacon and cook with the vegetables for a few minutes, then add the tomatoes, barley and stock; season.
3 Bring to the boil, cover and simmer for about 45 minutes.
4 Cool rapidly, pack and freeze.
5 To thaw: heat gently in a saucepan from frozen until boiling, adding the frozen peas and dumplings if you wish. Bring back to the boil and cook for about 15–20 minutes, until dumplings are light and well risen. (If only a few servings of the soup are needed, adjust peas and dumplings amounts accordingly.)

MAKES 1.8 litres ($3\frac{1}{4}$ pints)

Dumplings

200 g (7 oz) self-raising flour
pinch of salt
100 g (4 oz) shredded suet

1 Mix all ingredients to a dough with cold water.
2 Shape into ten small balls. Open freeze before packing.

MAKES 10 DUMPLINGS

Purée of Potato Soup

25 g (1 oz) butter
1 kg (2 lb) potatoes, peeled and roughly chopped
175 g (6 oz) roughly chopped onion
1.4 litres ($2\frac{1}{2}$ pints) unseasoned veal bone stock
10 ml (2 level tsp) salt
5 ml (1 level tsp) dried sage
freshly ground pepper

1 Melt the butter in a frying pan and fry the potatoes and onion without colouring them for about 5 minutes.
2 Add the remaining ingredients to the pan, bring to the boil, cover and simmer for 15–20 minutes.
3 Allow the soup to cool slightly, then rub through a sieve or purée in a blender.
4 Cool rapidly, pour into rigid containers and freeze.
5 To thaw: heat gently from frozen in a saucepan until boiling. Adjust seasoning and serve sprinkled with crisp, chopped bacon or croûtons.

MAKES 1.8 litres ($3\frac{1}{4}$ pints)

Chicken Soup

25 g (1 oz) butter
175 g (6 oz) finely chopped onion
1 litre ($1\frac{3}{4}$ pints) unseasoned chicken stock
about 200 g (7 oz) cooked chicken meat trimmings
568 ml (1 pint) milk
10 ml (2 level tsp) salt
2.5 ml ($\frac{1}{2}$ level tsp) freshly ground pepper
2.5 ml ($\frac{1}{2}$ level tsp) ground nutmeg
2.5 ml ($\frac{1}{2}$ level tsp) dried thyme
3 egg yolks, 142 ml ($\frac{1}{4}$ pint) single cream and chopped fresh parsley to serve

1 Melt the butter in a large saucepan and sauté the onion until soft.
2 Stir in all the remaining ingredients except the egg yolks, cream and parsley and bring to the boil.
3 Cover and simmer gently for 5 minutes.
4 Cool rapidly and freeze in rigid containers; seal and label.
5 To thaw: heat gently in a saucepan until boiling. Reduce the heat and stir in the egg yolks blended with the single cream. Heat gently without boiling, stirring, for 2–3 minutes. Adjust seasoning, serve sprinkled with chopped parsley. (When only a few servings of the soup are needed, adjust additional ingredients accordingly.)

MAKES 1.8 litres ($3\frac{1}{4}$ pints)

Duchesse and Croquette Potatoes

2.3 kg (5 lb) potatoes, peeled and roughly chopped
salt and freshly ground pepper
100 g (4 oz) butter
grated nutmeg (optional) and beaten egg for Duchesse Potatoes
50 g (2 oz) finely chopped onion, 1 egg white and about 175 g (6 oz) fresh breadcrumbs for Croquette Potatoes

1 Cook the potatoes in boiling salted water until tender. Drain well and mash until smooth, or put through a sieve.
2 Return to pan and stir in butter cut into pieces. Adjust seasoning with salt and pepper. Divide into two parts.
3 To make *Duchesse Potatoes*: Line two baking sheets with non-stick paper. With a piping bag fitted with a large star vegetable nozzle, pipe one portion of potato in whirls. If you wish, grate a little nutmeg over.

Open freeze until firm, remove from paper, store in a rigid container. To use, transfer as many frozen Duchesse Potatoes as needed to well-greased baking sheets and brush lightly with beaten egg. Place in cold oven, then bake at 200°C (400°F) mark 6 for 20–30 minutes until light golden brown.
4 To make *Croquette Potatoes*: To one portion of potato, stir in finely chopped onion, adjust seasoning and divide into sixteen equal parts. Roll each into a cork shape on a lightly floured work surface. Brush well with lightly beaten egg white and coat in breadcrumbs. Place on a baking sheet.

Open freeze until firm, store in polythene containers. To use, fry in deep fat at 190°C (375°F) for 8–10 minutes or use shallow frying pan, turning from time to time.

MAKES ABOUT 24 DUCHESSE POTATOES, 16 CROQUETTE POTATOES

Mushrooms à la Grecque

700 g ($1\frac{1}{2}$ lb) tomatoes, skinned and seeded
60 ml (4 tbsp) olive oil or corn oil
100 g (4 oz) finely chopped onion
2 garlic cloves, skinned and crushed
900 g (2 lb) button mushrooms, wiped
90 ml (6 tbsp) dry white wine
10 ml (2 level tsp) salt
freshly ground pepper

1 Chop the tomato flesh.
2 Heat the oil and sauté the onion gently until soft. Add the garlic, mushrooms, tomato, wine, salt and pepper.
3 Cook gently, uncovered, for about 10 minutes. Cool rapidly, pack and freeze.
4 To thaw: heat gently from frozen in a saucepan; serve hot or cold with crusty bread as an appetizer, garnished with chopped parsley.

MAKES 8 PORTIONS

Mushroom Duxelles

50 g (2 oz) butter
100 g (4 oz) finely chopped onion
225 g (8 oz) mushrooms, wiped and finely chopped
salt and freshly ground pepper

1 Melt the butter and gently fry the onion until soft. Add the mushrooms and continue cooking until the mixture is fairly dry. Season.
2 Turn into a small container, seal and freeze when cold.
3 Use from frozen to flavour sauces, omelettes, stuffings, etc.

CHICKEN DAY

Shopping List
Butcher **Six 1.6-kg (3½-lb) chickens, each jointed into 2 wings, 2 thighs, 2 drumsticks, 2 breasts, plus carcass and giblets; 225 g (8 oz) pork sausagemeat; 225 g (8 oz) streaky bacon rashers (optional).**
Greengrocer **175 g (6 oz) mushrooms, plus 225 g (8 oz) (optional); 225 g (8 oz) onions; 1 small red pepper; 1 small green pepper, plus another 225 g (8 oz) (optional); 1 lemon; 225 g (8 oz) baby onions (optional).**
Grocer **175 g (6 oz) Gruyère cheese; 2 eggs; 275 g (10 oz) butter.**
Store cupboard **Demerara sugar; French mustard; dry white wine; brandy; bouquet garni; dried thyme; plain flour; cornflour; oil; 175 g (6 oz) white bread for crumbs.**

Even today there are times when fresh chickens are cheaper, so watch the prices. More often than not, butchers are very willing to clean and joint the birds at no extra charge, but it's considerate to give them notice and say how you would like to have it done.

Cooking plan

1 Set freezer to fast-freeze; prepare fridge space for pre-chilling.
2 Remove skin from all joints of six chickens. Carefully cut breasts from bone.
3 Make stock from carcass, giblets (reserve livers) and skin, using enough water just to cover in one or two pans; simmer about 3 hours. (In a pressure cooker don't quite cover them with water, and cook in batches for about 45 minutes.)
4 Poach chicken wings for fricassée.
5 Bat out chicken breasts, coat in crumb mixture and open freeze.
6 Prepare and freeze chicken drumsticks.
7 Finish preparation of chicken fricassée and leave to cool in refrigerator.
8 Strain stock. (Remove any scraps of cooked chicken from bones as these can be frozen for boosting soups.) Chill.
9 In afternoon, prepare chicken casserole and chill ready for packing. Prepare and sauté separately the 225 g (8 oz) each of chopped bacon, chopped peppers, whole onions and sliced mushrooms. When cold, pack individually to add to casserole at serving time.
10 Prepare chicken liver pâté; pot, chill.
11 Pack open-frozen chicken breasts a few at a time in polybags.
12 Pack stock when cool, but not set. If it's not jellied enough, reduce by fast boiling and cool again. Pour some stock into ice trays and, when frozen, pack in polybags—small amounts are useful.
13 Pack chicken fricassée 'boil-in-bag' style and freeze.
14 Pack chicken casserole and freeze.
15 Pack and freeze pâté.

Chicken Liver Pâté

175 g (6 oz) butter
100 g (4 oz) chopped onion
6 chicken livers
30–45 ml (2–3 tbsp) brandy
45 ml (3 tbsp) chicken stock
225 g (8 oz) pork sausagemeat
2.5 ml (½ level tsp) dried thyme
salt and freshly ground pepper
1 lemon

1 Melt 50 g (2 oz) butter in a frying pan and gently sauté the onion and chicken livers for 5 minutes. Place in a blender with the brandy and stock, and blend until smooth.
2 Fry sausagemeat gently for a few minutes until broken down, add puréed liver, thyme and seasoning to taste. Stir over a moderate heat for 10 minutes.
3 Pack into individual pots, level the surface, place a slice of lemon on each and seal with melted butter. Chill. Overwrap individually.
4 To use, thaw at room temperature for about 1½ hours or overnight in the fridge.

MAKES ENOUGH FOR 6–8

Basic Chicken Casserole

25 g (1 oz) butter
15 ml (1 tbsp) oil
12 chicken thighs
50 g (2 oz) plain flour
1 litre ($1\frac{3}{4}$ pints) chicken stock
salt and freshly ground pepper
225 g (8 oz) chopped bacon and 225 g (8 oz) baby onions, sautéed; or 225 g (8 oz) sliced mushrooms and 225 g (8 oz) chopped green pepper, sautéed, to serve

1 Heat the butter and oil in a 2.3-litre (4-pint) flameproof casserole, and fry the chicken on all sides until golden. Remove the chicken and keep hot.
2 Stir the flour into the fat in the casserole and cook gently for 2 minutes. Gradually add the stock and bring to the boil, stirring.
3 Add the chicken, and salt and pepper to taste; cover and simmer gently for about 30 minutes, until the chicken is tender. Skim off excess fat.
4 Cool rapidly and divide equally between two containers. Cover and freeze.
5 To use, unwrap one pack, place in a casserole and cook, covered, from frozen in the oven at 190°C (375°F) mark 5 for about 30 minutes until softened, then stir in either sautéed bacon and baby onions (this will serve 4) or sautéed mushrooms and peppers (enough for 3). Cook a further 30 minutes.

MAKES 7 PORTIONS

Chicken and Pepper Fricassée

12 chicken wings
150 ml ($\frac{1}{4}$ pint) dry white wine
5 ml (1 level tsp) salt
1 bouquet garni
10 black peppercorns
100 g (4 oz) onion, skinned and finely chopped
1 small red pepper, seeded and finely chopped
1 small green pepper, seeded and finely chopped
25 g (1 oz) butter
30 ml (2 level tbsp) cornflour

1 Place the chicken wings in a pan with 300 ml ($\frac{1}{2}$ pint) water, wine, salt, bouquet garni and peppercorns.
2 Bring to the boil, cover and simmer gently for about 30 minutes, until the chicken is just tender.
3 Strain, skim and reserve the liquid, and trim all the meat from the chicken bones.
4 In a saucepan, sauté the onion and peppers gently in the butter until tender.
5 Blend the cornflour with a little of the reserved liquid, then stir in the rest and add to the sautéed vegetables.
6 Bring to the boil and simmer for 3–4 minutes. Add the chicken meat, adjust seasoning, and cool. Pack 'boil-in-the-bag' style and freeze.
7 To use, place frozen boil-in-bag in boiling water and simmer until thawed and reheated—about 45 minutes. Serve with rice or pasta.

MAKES 4 PORTIONS

Crumbed Chicken Escalope

175 g (6 oz) fresh breadcrumbs
175 g (6 oz) mushrooms, wiped and finely chopped
175 g (6 oz) Gruyère cheese, finely grated
salt and freshly ground pepper
2 eggs, beaten
12 chicken breasts

1 Combine breadcrumbs, mushrooms and cheese in a large bowl. Season lightly.
2 Dip chicken breasts into beaten egg then coat in breadcrumb mixture; pat crumbs well in to ensure that each breast is evenly coated.
3 Open freeze on baking sheets, then freeze in polythene bags.
4 To use, shallow fry from frozen in oil or lard for about 15 minutes, turning once.

MAKES 12 PORTIONS

Devilled Drumsticks

50 g (2 oz) butter
15 ml (1 level tbsp) demerara sugar
10 ml (2 level tsp) French mustard
2.5 ml ($\frac{1}{2}$ level tsp) salt
12 chicken drumsticks

1 Melt the butter in a saucepan and add the sugar, mustard and salt.
2 Place the scored drumsticks in a mixing bowl and pour over the butter mixture, turning the drumsticks to coat evenly.
3 When butter glaze has set, wrap each drumstick separately in foil. Overwrap in a polybag. Freeze.
4 To use, remove drumsticks from bag and, still in foil, place on a baking sheet. Cook from frozen in the oven at 190°C (375°F) mark 5 for $1\frac{1}{4}$ hours; open up the foil for the last 20 minutes to brown.

MAKES 6 PORTIONS

CAKE AND COOKIE DAY

Greengrocer **2 lemons.**
Grocer **2 kg ($4\frac{1}{2}$ lb) butter or block margarine (or half and half if you like); 2 kg ($4\frac{1}{2}$ lb) caster sugar; 450 g (1 lb) icing sugar; 1.3-kg (3-lb) bag each of self-raising and plain flour; 24 eggs; 50 g (2 oz) plain chocolate; 50 g (2 oz) mixed dried fruit; 175 g (6 oz) currants; 175 g (6 oz) sultanas; 50 g (2 oz) glacé cherries; 50 g (2 oz) blanched almonds.**
Store cupboard **50 g (2 oz) marzipan; cocoa powder; coffee essence; almond and vanilla flavourings; mixed spice; ground ginger; clear honey; apricot jam; orange marmalade; milk; lemon curd; angelica; desiccated coconut; crystallized or stem ginger.**

Cook-ahead baking sessions are among the most profitable and rewarding. If you can't bake all the mixtures we've suggested, freeze what's uncooked in the correctly weighed-off amounts, and then thaw and bake as needed. Uncooked, creamed mixtures are splendid standbys anyway and benefit from their 'rest' in the freezer.

We've planned this session so it can be split easily into half days—one for cakes, one for cookies. You'll need plenty of working space and racks for cooling. To keep the oven working to capacity, duplicate baking tins, too. Cooking times and temperatures are for a full oven. When only one or two cakes are baked, reduce temperature by one mark, or by 50°C (122°F).

Cooking plan

1 In the morning, put freezer switch on fast-freeze and check space available. Take refrigerated fat and eggs out to reach room temperature. Get store cupboard items out on tray; light oven.
2 Prepare and bake Madeira and Fruit Cakes on oven's bottom rung.
3 Prepare and bake the Chocolate and Vanilla Slabs at oven centre or just above, along with Madeira and Fruit. If necessary, bake in rotation. Make and bake the Spice Slab variation.
4 Meanwhile, prepare mixture for puddings and small cakes, baking latter when oven is free. Freeze puddings.
5 In the afternoon, make and bake the batch of cookies. Prepare butter icing.
6 Pipe coffee butter icing with a small rose vegetable nozzle on coffee buns. Roll out almond paste and use to decorate small almond cakes with angelica.
7 Cut up, sandwich Chocolate Slabs with vanilla butter icing, and decorate.
8 Open freeze Coffee Cakes and Chocolate Slices.
9 Pair chocolate biscuits with remaining butter icing.
10 Pack, label and freeze all cakes and cookies, when cool, on day of baking.

Butter Icing

450 g (1 lb) butter, or 225 g (8 oz) butter and 225 g (8 oz) block margarine
450 g (1 lb) icing sugar
coffee essence
vanilla flavouring

1 Beat the butter in a large bowl until creamy. Add the sifted icing sugar a little at a time, beating well between each addition.
2 Weigh off 175 g (6 oz) into a small bowl and flavour with coffee essence to taste.
3 Flavour remainder with vanilla.

MAKES ABOUT 900 g (2 lb)

Madeira and Mixed Fruit Cakes

350 g (12 oz) butter or block margarine
350 g (12 oz) caster sugar
6 eggs
275 g (10 oz) self-raising flour
225 g (8 oz) plain flour
1 lemon
4 thin slices of lemon peel
175 g (6 oz) currants
175 g (6 oz) sultanas
50 g (2 oz) glacé cherries, quartered
2.5 ml ($\frac{1}{2}$ level tsp) mixed spice

1 Grease and line two 18-cm (7-inch) round cake tins.
2 Beat butter and sugar together in a bowl until light and fluffy. Gradually beat in eggs, then fold in sifted flours using a metal spoon. Divide mixture into two equal parts.
3 For Madeira Cake: fold into one part the grated rind and juice of $\frac{1}{2}$ lemon, turn mixture into one prepared cake tin, level the surface and place two slices of lemon peel on top. Bake on bottom oven shelf at 190°C (375°F) mark 5 about $1\frac{1}{2}$ hours.
4 For Rich Fruit Cake: fold in remaining ingredients with grated rind of other half of lemon, remaining lemon peel, chopped and turn into the other tin. Bake alongside the first for about $1\frac{1}{2}$ hours. Cool on wire rack. Pack and freeze.
5 To use, thaw, wrapped, at room temperature about 4 hours.

Coffee/Almond Small Cakes and Three Sponge Puddings

350 g (12 oz) butter or block margarine
350 g (12 oz) caster sugar
6 eggs
400 g (14 oz) self-raising flour
50 g (2 oz) mixed dried fruit
30 ml (2 level tbsp) orange marmalade
finely grated rind of 1 lemon
30 ml (2 level tbsp) lemon curd
2.5 ml ($\frac{1}{2}$ tsp) almond flavouring
2.5 ml ($\frac{1}{2}$ tsp) coffee essence
175 g (6 oz) coffee butter cream, made weight
50 g (2 oz) marzipan, apricot jam and angelica to decorate

1 Grease three 450-g (1-lb) foil pudding basins.
2 Cream fat and sugar until light and fluffy. Gradually beat in eggs, then lightly beat in flour. Weigh out 850 g (1 lb 14 oz) of mixture.
3 Divide measured quantity into three and stir fruit into first, marmalade into second; lemon rind into third. Put lemon curd in one basin and cover with lemon mixture; other flavours go in other basins. Seal, label and freeze puddings.
4 Divide remaining mixture into two. Stir almond flavouring into one, coffee essence into other.
5 Put thirty-six paper cases into patty tins (keeps their shape). Divide mixtures evenly between cases; eighteen of each. Bake just above the oven centre at 190°C (375°F) mark 5 for about 20 minutes. Cool on wire rack.
6 Pipe butter cream on to coffee buns; brush tops of almond buns with apricot jam, decorate with star shapes cut from thinly rolled almond paste and angelica. Pack in rigid box; freeze.
7 To use, steam puddings from frozen about $2\frac{1}{4}$ hours. Thaw small cakes unwrapped.

MAKES 36 CAKES AND 3 PUDDINGS

Layered Chocolate Slices and Vanilla Slab

450 g (1 lb) butter or block margarine
450 g (1 lb) caster sugar
8 eggs
450 g (1 lb) self-raising flour
5 ml (1 tsp) vanilla flavouring
45 ml (4 level tbsp) cocoa powder
30 ml (3 tbsp) milk
575 g (21 oz) vanilla butter cream, made weight
50 g (2 oz) chocolate, grated

1 Grease and base-line three 28 × 18 × 3 cm (11 × 7 × 1¼ inches) cake tins (or see **3**).
2 Beat fat until soft, add sugar, and cream until light and fluffy. Beat in the eggs, a little at a time, then lightly beat in flour and vanilla.
3 Take out one-third of mixture and put in one prepared tin (or divide between two 18-cm (7-inch) greased and base-lined sandwich tins).
4 Blend cocoa with milk; gently beat into the remaining mixture. Divide between the remaining two tins.
5 Bake all the cakes in the centre of the oven or just above at 190°C (375°F) mark 5 for about 25 minutes.
6 Turn out, cool on wire rack. Halve each lengthwise; wrap plain cakes in cling film or foil; freeze for filling later. Sandwich the chocolate strips in pairs with vanilla butter cream to make two cakes. Zig-zag piped butter cream and grated chocolate down centres. Open freeze until firm, then pack.
7 To use, unwrap chocolate slices and thaw at room temperature about 2 hours. Thaw plain slabs, wrapped, about 1½ hours; finish as preferred.

MAKES 3 CAKES

Honey Spice variation Make as above but omit vanilla flavouring, cocoa and milk. Add 20 ml (4 level tsp) mixed spice and 90 ml (6 tbsp) clear honey. Divide between the three oblong tins. Scatter 100 g (4 oz) chopped walnuts over two tins. Bake; then cool, wrap, freeze and thaw as above. Plain slab may be glacé-iced.

Cookie Medley

Basic recipe
450 g (1 lb) butter or block margarine
550 g (1 lb 4 oz) caster sugar
4 egg yolks

Cream together fat and sugar until fluffy; beat in egg yolks, divide into four portions and add the following variations:

Ginger Biscuits

225 g (8 oz) plain flour
5 ml (1 level tsp) ground ginger
25 g (1 oz) crystallized or stem ginger, finely chopped

1 Stir sifted flour and ground ginger with crystallized ginger into one part biscuit mixture; knead well. Roll out 1.25-cm (¼-inch) thick; cut out twenty-four biscuits with a 7-cm (2¾-inch) long oval cutter: place on greased baking sheets.
2 Bake in the oven at 180°C (350°F) mark 4 for about 15 minutes, until golden. Cool on a wire rack; pack and freeze.
3 To use: unwrap and place on baking sheet, refresh from frozen at 200°C (400°F) mark 6 for about 5 minutes. Dredge with caster sugar, cool on a wire rack.

MAKES 24

Chocolate Creams

200 g (7 oz) plain flour
25 g (1 oz) cocoa powder
150 g (5 oz) vanilla butter cream, made weight

1 Stir sifted flour and cocoa into one part biscuit mixture, knead well. Roll out 1.25-cm ($\frac{1}{4}$-inch) thick; cut out twenty-four biscuits using 5-cm (2-inch) round fluted cutter; place on greased baking sheets.
2 Bake in the oven at 180°C (350°F) mark 4 for 15–20 minutes. Cool on a wire rack.
3 Sandwich biscuits together in pairs with vanilla butter cream; pack and freeze.
4 To use, thaw at room temperature for about 30 minutes, sprinkle with sifted icing sugar.

MAKES 12

Almond Bars

200 g (7 oz) plain flour
50 g (2 oz) finely chopped almonds
2.5 ml ($\frac{1}{2}$ tsp) almond flavouring

1 Stir flour, 25 g (1 oz) almonds and flavouring into one part biscuit mixture; press into greased 18-cm (7-inch) square shallow tin, prick well and sprinkle with remaining almonds.
2 Bake in the oven at 180°C (350°F) mark 4 for 20–25 minutes. While still warm, cut into sixteen bars; when cold, pack and freeze.
3 To use, unwrap and place on a baking sheet, refresh from frozen at 200°C (400°F) mark 6 for 5 minutes. Dredge with caster sugar and cool on a wire rack.

MAKES 16

Coconut Cookies

175 g (6 oz) plain flour
50 g (2 oz) desiccated coconut

1 Stir flour and coconut into one part biscuit mixture; knead well. Form into twenty-four balls and place on well-greased baking sheet. Press each lightly with dampened fork.
2 Bake in the oven at 180°C (350°F) mark 4 for 15–20 minutes. Remove from baking tray before completely cold, cool on wire rack; pack and freeze.
3 To use, unwrap and place on a baking sheet, refresh from frozen at 200°C (400°F) mark 6 for 5 minutes. Dredge with caster sugar and cool on a wire rack.

MAKES 24

Suppers from the freezer

Tomato-Topped Beef Loaf

450 g (1 lb) boned top rib of beef
125 g (4 oz) onions, skinned
50 g (2 oz) fresh breadcrumbs
1 lemon
1 garlic clove, skinned and crushed
5 ml (1 level tsp) dried basil
2.5 ml ($\frac{1}{2}$ level tsp) salt
freshly ground pepper
5 ml (1 level tsp) beef extract
1 egg, size 2
225 g (8 oz) tomatoes, skinned, seeded and roughly chopped
60 ml (4 level tbsp) mango chutney
6 stuffed green olives, sliced

1 Mince beef and onion together, then combine with breadcrumbs, finely grated lemon rind, 15 ml (1 tbsp) lemon juice, garlic, basil and seasoning. Stir in beef extract and egg, beating well to blend.
2 Mix tomatoes with the chutney. Spoon into a greased 1.1-litre (2-pint) foil-lined loaf tin. Scatter sliced olives over tomato mixture. Fill the tin with the meat mixture and cover tightly with greased foil.

To freeze

Freeze in tin. When frozen, remove from tin and overwrap. To use, unwrap, replace in loaf tin, stand the tin in a water bath and bake from frozen at 180°C (350°F) mark 4 for about $2\frac{1}{2}$ hours. Cover the top with foil if it is becoming too brown.

MAKES 4 PORTIONS

Wholewheat Pizza

15 g ($\frac{1}{2}$ oz) fresh yeast, or 7.5 ml ($1\frac{1}{2}$ level tsp) dried yeast
120 ml (4 fl oz) tepid milk and water, mixed
5 ml (1 level tsp) sugar
100 g (4 oz) plain flour
100 g (4 oz) wholewheat flour
7.5 ml ($1\frac{1}{2}$ level tsp) salt
90 ml (6 tbsp) oil
15 ml (1 tbsp) French mustard
350 g (12 oz) tomatoes, skinned and sliced
75 g (3 oz) mushrooms, wiped and sliced
10 ml (2 tsp) chopped fresh chives
salt and freshly ground pepper
125 g (4 oz) grated cheese—Mozarella, or a mixture of Edam, Gruyère, Cheddar and blue cheese

1 Cream the fresh yeast with a little of the liquid and add remainder. If using dried yeast, dissolve the sugar in all the liquid then sprinkle the yeast over the surface; leave to stand in a warm place for about 20 minutes, until really frothy.
2 Sift the flour and salt into a bowl, stir in sugar (for fresh yeast). Make a well in the centre, pour in the yeast liquid and 30 ml (2 tbsp) oil and gradually combine to make a soft dough. Knead well on a floured work surface for at least 5 minutes.
3 Leave to rise, covered, for about 1 hour, until double in size. Knead again and divide into two pieces. Shape into balls and leave, covered, for about 15 minutes.
4 Roll out each portion of dough on an oiled baking sheet to a 23-cm (9-inch) round. Crimp up edges to give a rim. Brush each base with 15 ml (1 tbsp) oil. Spread with mustard, top with tomatoes, mushrooms, chives, seasoning and cheese. Sprinkle with the remaining oil.
5 Bake in the oven at 230°C (450°F) mark 8 for 15–18 minutes, until well browned and crisp. Cool quickly.

To freeze

Open freeze until firm, then store in freezer bags. To use, place on baking sheets and thaw for about 1 hour. Cover loosely with foil and reheat in the oven at 180°C (350°F) mark 4 for about 20 minutes. Remove foil and heat for a further 10 minutes.

MAKES 2 PORTIONS

Barbecued Riblets of Lamb

2 large breasts of lamb
30 ml (2 tbsp) vinegar
10 ml (2 level tsp) ground ginger
45 ml (3 tbsp) soy sauce
45 ml (3 tbsp) clear honey
60 ml (4 level tbsp) ginger marmalade
15 ml (1 tbsp) Worcestershire sauce
226-g (8-oz) can tomatoes, with juice
30 ml (2 tbsp) lemon juice
salt and freshly ground pepper

1 Trim any excess fat off the lamb and cut down between the bones to form riblets.
2 Place the meat in a large saucepan, cover with cold water and add the vinegar. Bring to the boil and simmer for 10 minutes, then drain well in a colander. Place the meat in a foil tin.
3 Meanwhile gently heat the remaining ingredients in a small pan until blended. Pour over the meat.
4 Bake in the oven at 190°C (375°F) mark 5 for 50–60 minutes, turning and basting frequently. Cool.

To freeze
When cold, pack and freeze. To use, thaw in a cool place overnight, spoon into a roasting tin, cover with kitchen foil, and cook at 180°C (350°F) mark 4 for about 35 minutes. Serve with thick slices of warm French bread and a salad of chicory, orange and watercress.

MAKES 4 PORTIONS

Chilli Con Carne

225 g (8 oz) dried red kidney beans
225 g (8 oz) onions, skinned and chopped
15 ml (1 level tbsp) oil
700 g (1½ lb) lean minced beef
1 large garlic clove, skinned and crushed
15 ml (1 level tbsp) salt
freshly ground pepper
30–45 ml (2–3 level tbsp) chilli seasoning, or 2.5 ml (½ level tsp) chilli powder
15 ml (1 level tbsp) plain flour
30 ml (2 level tbsp) tomato purée
793-g (28-oz) can tomatoes, with juice

1 Soak the beans overnight, then boil gently in unsalted water for about 45 minutes, until tender.
2 In a large saucepan, fry the onion in the oil, add mince and cook until brown and crumbly. Add the crushed garlic to mince with the salt, pepper and chilli seasoning. Sprinkle on the flour and stir. Add tomato purée and the tomatoes with their juice.
3 Bring to the boil and add drained beans. Reduce heat and simmer for 30 minutes, stirring occasionally.

To freeze
Cool and place in freezer container. To use, reheat from frozen in a covered casserole at 180°C (350°F) mark 4 for about 45 minutes. Serve with rice.

MAKES 6 PORTIONS

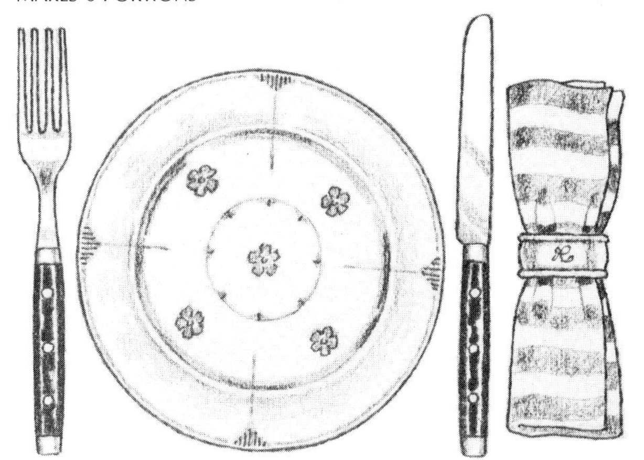

Baked Bean and Beef Lasagne

175 g (6 oz) onions, skinned and chopped
30 ml (2 tbsp) vegetable oil
450 g (1 lb) lean minced beef
10 ml (2 level tsp) ground paprika
450 ml (¾ pint) beef stock
396-g (14-oz) can tomatoes, with juice
450-g (1 lb) can baked beans
1 garlic clove, skinned and crushed
salt and freshly ground pepper
225 g (8 oz) lasagne
40 g (1½ oz) butter or block margarine
30 ml (2 level tbsp) plain flour
568 ml (1 pint) milk
175 g (6 oz) Edam cheese, grated

1 In a frying pan, fry the onion until golden in the oil. Add the mince and brown well. Stir in the paprika, stock, tomatoes with juice, baked beans, crushed garlic and seasoning. Simmer, covered for 20 minutes.
2 Place a layer of the bean-mince mixture in a 2.3-litre (4-pint) 30.5 × 20.5 × 6.5-cm (12 × 8 × 2½-inch) dish, then cover with uncooked lasagne. Continue layering until all the mince and lasagne is used up, ending with a layer of mince.
3 Make a cheese sauce in the ordinary way from the fat, flour, milk, seasoning and half the grated cheese. Spoon over the mince mixture. Sprinkle remaining cheese over the top then bake, uncovered, in the oven at 180°C (350°F) mark 4 for about 50 minutes.

To freeze
Cool, pack and freeze in dish. To use, thaw at cool room temperature overnight, then reheat in the oven at 180°C (350°F) mark 4 for about 1 hour 20 minutes.

MAKES 6 PORTIONS

Crêpes Ratatouille

275 g (10 oz) aubergine, cut into 2.5-cm (1-inch) cubes
25 g (1 oz) butter or block margarine
225 g (8 oz) courgettes, thickly sliced
225 g (8 oz) onions, skinned and roughly chopped
225 g (8 oz) tomatoes, skinned and roughly chopped
1 garlic clove, skinned and crushed
5 ml (1 level tsp) dried oregano
225 g (8 oz) salami, sliced and shredded
salt and freshly ground pepper
300 ml (½ pint) pancake batter
150 ml (¼ pint) dry white wine
45 ml (3 level tbsp) chopped fresh parsley

1 Sprinkle the aubergine cubes liberally with salt. Leave for 1 hour, then rinse well.
2 Melt the butter or margarine in a deep frying pan. Add the aubergines, courgettes, onion, tomatoes, garlic and 2.5 ml (½ level tsp) oregano. Cook gently, covered, for 25–30 minutes, until the vegetables are tender. Add the salami and season.
3 Meanwhile make six 20.5-cm (8-inch) pancakes from the batter and place on a work surface. With a slotted spoon divide the mixture between the pancakes. Roll up and place in a buttered ovenproof or foil dish. Stir the wine into the pan juices with the parsley and the remaining oregano. Season. Pour over the pancakes.

To freeze
Cool, then pack and freeze in dish. To use, thaw for 2 hours in cool room, then cover with foil and reheat in the oven at 190°C (375°F) mark 5 for about 25 minutes.

MAKES 6 PORTIONS

Casseroles for entertaining

Pork and Bacon Casserole

215-g (7½-oz) packet puff pastry, thawed
1 egg, beaten
1.1 kg (2½ lb) boneless chump end pork or pork fillet
30 ml (2 tbsp) vegetable oil
50 g (2 oz) butter
100 g (4 oz) streaky bacon, cut into small pieces
350 g (12 oz) baby onions, skinned
60 ml (4 level tbsp) flour
600 ml (1 pint) stock
finely grated rind ½ lemon
30 ml (2 level tbsp) chopped fresh sage, or 5 ml (1 level tsp) dried sage
salt and freshly ground pepper
45 ml (3 tbsp) medium sherry
chopped fresh parsley to serve

1 Roll out the pastry to 0.5-cm (¼-inch) thick and stamp out crescents using a 6.5-cm (2½-inch) round cutter. Place on a greased baking sheet, glaze with beaten egg and bake in the oven at 220°C (425°F) mark 7 for about 15 minutes. Cool on a wire rack.
2 Cut up the pork into large bite-size pieces, discarding excess fat. Heat the oil and butter in a flameproof casserole and brown the pork and bacon well, a little at a time. Drain.
3 Add the onions to the remaining oil and brown lightly; mix in the flour and fry gently for 1–2 minutes. Add the stock with the lemon rind, herbs and seasoning and bring to the boil, stirring. Return the meat to the pan and cook, covered, in the oven at 150°C (300°F) mark 2 for about 2 hours.

To freeze
Cool, pack in preformer and freeze; pack the fleurons (the pastry crescents) separately. To use, thaw the casserole at cool room temperature overnight, bring to the boil on top of the cooker and simmer gently, covered, for 10 minutes. Add the sherry and garnish with parsley. To thaw fleurons, place still frozen, on a baking sheet, cover loosely with foil and refresh in the oven at 220°C (425°F) mark 7 for 10 minutes.

MAKES 6 PORTIONS

Osso Buco with Tomato

6 osso buco, about 225 g (8 oz) each
30 ml (2 level tbsp) plain flour
salt and freshly ground pepper
400 ml (¾ pint) dry white wine
30 ml (2 tbsp) vegetable oil
25 g (1 oz) butter
225 g (8 oz) onions, skinned and sliced
450 g (1 lb) tomatoes, skinned and roughly chopped
30 ml (2 level tbsp) tomato purée
1 garlic clove, skinned
2.5 ml (½ level tsp) dried basil
chopped fresh parsley

1 Trim the osso buco (cut from knuckle of veal) to remove any excess fat. Season the flour well and use to lightly coat the meat. In a small pan boil down the wine to reduce by half.
2 Heat the oil and butter in a large shallow flameproof dish and brown the meat well on both sides. Lift the meat out of the pan; lightly brown the onion in the remaining fat. Add the tomatoes, wine, tomato purée, crushed garlic, basil and more seasoning.
3 Bring to the boil, replace the meat and cover the dish tightly. Cook in the oven at 170°C (325°F) mark 3 for 1¾–2 hours, until the meat is tender.

To freeze
Cool, pack and freeze. To use, thaw at cool room temperature overnight. Place in a casserole and reheat in the oven covered at 200°C (400°F) mark 6 for about 40 minutes. Lift the meat out of the cooking juices and keep warm. Reduce the cooking juices for 2–3 minutes to thicken slightly. Adjust seasoning, then spoon the sauce over the veal. Sprinkle with chopped parsley.

MAKES 6 PORTIONS

Eastern Casserole of Lamb

1.4 kg (3 lb) lean boned lamb or mutton
450 g (1 lb) tomatoes, skinned
45 ml (3 tbsp) oil
350 g (12 oz) onions, skinned and sliced
30 ml (2 level tbsp) ground coriander
10 ml (2 level tsp) ground cumin
5 ml (1 level tsp) ground ginger
1.25 ml ($\frac{1}{4}$ level tsp) turmeric
1.25 ml ($\frac{1}{4}$ level tsp) cayenne pepper
30 ml (2 level tbsp) plain flour
15 ml (1 level tbsp) tomato purée
450 ml ($\frac{3}{4}$ pint) chicken stock
75 g (3 oz) sultanas
salt and freshly ground pepper

1 Cut the meat into largish chunks. Quarter the tomatoes; push the seeds through a sieve and keep the juice.
2 Heat the oil in a large flameproof casserole and brown the meat well, a few pieces at a time, then remove from casserole. Brown the onions in the fat remaining in the casserole. Add the coriander, cumin, ginger, turmeric and cayenne and cook for 2 minutes, then stir in the flour and cook for a further 2 minutes. Add the tomato purée, stock and sultanas and season well.
3 Return the meat to the casserole, cover, and cook in the oven at 170°C (325°F) mark 3 for about 1$\frac{1}{4}$ hours for lamb, 2$\frac{1}{2}$ hours for mutton.

To freeze
Cool, pack in two shallow portions and freeze. To use, thaw overnight at cool room temperature. Reheat on the top of the cooker.

MAKES 8 PORTIONS

Lamb with Apricots

50 g (2 oz) dried apricots
125 g (4 oz) dried black-eye beans
550 g (1$\frac{1}{4}$ lb) lean boned lamb, cut into 2.5-cm (1-inch) cubes
30 ml (2 level tbsp) plain flour
2.5 ml ($\frac{1}{2}$ level tsp) chilli powder
15 ml (1 level tbsp) ground coriander
salt and freshly ground pepper
15 ml (1 tbsp) oil
125 g (4 oz) onions, skinned and sliced
125 g (4 oz) mushrooms, wiped and sliced
600 ml (1 pint) chicken stock
45 ml (3 level tbsp) mango chutney
142 g (5 oz) natural yoghurt
chopped fresh parsley to garnish

1 Soak the apricots and beans separately overnight. Boil beans for 10 minutes in unsalted water; drain.
2 Toss the meat in flour seasoned with chilli, coriander, salt and pepper, then heat the oil in a flameproof casserole and brown the meat, a few pieces at a time. Remove from the fat, using a draining spoon. Add the onion and mushrooms to the pan and fry for 2–3 minutes. Stir in the stock, beans, chutney, drained apricots.
3 Return the lamb to the casserole. Bring to the boil, cover and cook in the oven at 170°C (325°F) mark 3 for about 1 hour, until the meat is quite tender. Stir in the yoghurt, adjust seasoning.

To freeze
Cool, pack in preformer and freeze. To use, thaw overnight at cool room temperature, reheat gently on top of the stove. Garnish with parsley.

MAKES 4 PORTIONS

West African Beef Curry

2 kg (4 lb) chuck steak
50 g (2 oz) plain flour
1.25 ml ($\frac{1}{4}$ level tsp) ground paprika
1.25 ml ($\frac{1}{4}$ level tsp) cayenne pepper
1.25 ml ($\frac{1}{4}$ level tsp) chilli powder
45 ml (3 tbsp) corn oil
450 g (1 lb) onions, skinned and chopped
30 ml (2 level tbsp) desiccated coconut
90 ml (6 level tbsp) curry paste
1 garlic clove, skinned and crushed
few drops Tabasco sauce
1.1 litre (2 pints) beef stock

1 Trim and cut the steak into fork size pieces. Toss in flour seasoned with paprika, cayenne and chilli powder; use just enough flour to coat.
2 Heat oil in a large pan and fry the onions until well browned. Add the coconut, curry paste, garlic, Tabasco and stock.
3 In a large pan heat enough oil to just cover base, fry the meat, a little at a time, until browned. Drain and add it to the curry sauce. Cover tightly and simmer on top of the cooker for about 2 hours, until the meat is tender.

To freeze

Cool and pack in two shallow portions. To use, place in 1 large or 2 small pans with a little stock and heat gently.

MAKES 8 PORTIONS

Salmis of Pheasant

a brace of young pheasants
75 g (3 oz) butter
salt and freshly ground pepper
300 ml ($\frac{1}{2}$ pint) chicken stock
slices of carrots, onion
bay leaf
225 g (8 oz) carrots, peeled and diced
125 g (4 oz) celery, cleaned and diced
125 g (4 oz) onions, skinned and chopped
45 ml (3 level tbsp) plain flour
60 ml (4 tbsp) port, or red wine reduced by half

1 Rinse the pheasants and dry well. Place a small knob of butter inside each bird and spread 25 g (1 oz) butter over the breasts; season well. Place in a roasting tin and pour over the chicken stock. Roast in the oven at 200°C (400°F) mark 6 for 40 minutes, basting occasionally. Reserve juices and joint pheasants.
2 Place backbones in a small pan. Pour over reserved roasting juices. Add the sliced carrot and onion with the bay leaf, seasoning and sufficient water to just cover the bones. Bring to the boil, skim, and simmer uncovered for about 30 minutes, until about 400 ml ($\frac{3}{4}$ pint) stock remains. Strain and set aside.
3 Melt remaining butter in a flameproof casserole. Add carrots, celery and onions and sauté, covered for about 10 minutes. Add flour and cook 1 minute stirring. Blend in the reserved stock and port. Season, bring to boil, then remove from the heat. Place pheasant joints among the vegetables. The legs take longest to cook so put these at the bottom. Baste with the strained sauce. Cover. Place a piece of non-stick paper under the lid and cook in the oven at 170°C (325°F) mark 3 for about 1$\frac{1}{4}$ hours.

To freeze

Cool, pack in preformer and freeze. To use, return to casserole, thaw overnight or cool room temperature. Reheat at 180°C (350°F) mark 4 for about 45 minutes. Serve with croûtons.

MAKES 4 PORTIONS

Puddings that wait

Chocolate Supreme

225 g (8 oz) plain chocolate
175 g (6 oz) madeira-type cake
100 g (4 oz) ground almonds
60 ml (4 tbsp) orange-flavoured liqueur
75 ml (5 level tbsp) apricot jam, warmed
50 g (2 oz) unsalted butter
105 ml (7 level tbsp) icing sugar, sifted
90 ml (6 tbsp) double cream

1 Line the base of an 18.5-cm ($7\frac{1}{2}$-inch) loose-based fluted flan tin with non-stick paper. Over hot water, melt 175 g (6 oz) chocolate in a bowl and, using a pastry brush, coat the base and sides of the tin in layers—refrigerating between each layer to set the chocolate. Refrigerate completed shell.
2 In a bowl, crumble the cake and mix with the almonds. Stir in the liqueur and jam, then gently press into the chocolate shell. Refrigerate for at least 2 hours.
3 With an electric hand mixer beat the butter until creamy, then beat in the sugar and remaining chocolate, melted. Gradually beat in the cream.
4 Briefly dip the flan tin in hot water to loosen the shell. Transfer flan to a piece of kitchen foil. Pipe the chocolate cream over the truffle filling.

To freeze
Open freeze then overwrap. To use, thaw overnight in the refrigerator. Serve chilled.

MAKES 8 PORTIONS

Blackcurrant and Apple Summer Pudding

450 g (1 lb) fresh blackcurrants
450 g (1 lb) dessert apples, peeled, cored and roughly chopped
60 ml (4 level tbsp) redcurrant jelly
2.5 ml ($\frac{1}{2}$ level tsp) ground cinnamon
125 g (4 oz) demerara sugar
10 thin slices bread, crusts removed

1 Strip the blackcurrants of their stalks, rinse and drain. Place with the apples in a saucepan with the redcurrant jelly, cinnamon and 400 ml ($\frac{3}{4}$ pint) water. Bring slowly to the boil, cover the pan and simmer for 15–20 minutes, until the apples are quite soft. Remove from the heat, add the sugar and stir until dissolved.
2 Completely line a 1.4-litre ($2\frac{1}{2}$-pint) foil-lined pudding basin or loaf tin with the bread slices, reserving some for the top.
3 Spoon the blackcurrant mixture into the lined basin, reserving about 150 ml ($\frac{1}{4}$ pint) of the juices. Top with the remaining bread slices. Place a plate which is slightly smaller than the basin top on the pudding surface and weight down lightly. As the juices ooze into the bread take off the weights and spoon over reserved juices. Replace the weights and refrigerate for at least 24 hours.

To freeze
Remove the weights and leave the pudding in the container. Overwrap and freeze. To use, thaw at cool room temperature overnight. Turn out and serve chilled, with lightly whipped cream.

MAKES 6 PORTIONS

Gooseberry Mallow Ice Cream

12 white marshmallows
170-g (6-oz) can evaporated milk
450 g (1 lb) fresh or frozen gooseberries, prepared
75 g (3 oz) caster sugar
150 ml ($\frac{1}{4}$ pint) double cream
50 g (2 oz) golden syrup
12–16 macaroons to serve

1 Place the marshmallows and evaporated milk in a small bowl and then stand the bowl in a pan of warm water to melt the marshmallows. Stir until smooth, then cool.
2 Place half the gooseberries in a pan with 30 ml (2 tbsp) water. Cover and cook gently for about 5 minutes, until the skins burst and the fruit softens.
3 Stir the sugar into the warm fruit then rub through a sieve or purée in a blender. Cool.
4 Lightly whip the cream and stir into the marshmallow mixture with the fruit purée.
5 Spoon into a deep freezer-proof container and freeze until required. Do not churn.
6 Meanwhile, prepare the sauce: place remaining gooseberries with the golden syrup and 30 ml (2 tbsp) water in a small saucepan. Cover, cook gently until gooseberries are soft. Rub through a sieve or purée in a blender.

To freeze
Freeze as ice cream (see page 61) and sauce separately. To use, leave the ice cream to soften in the refrigerator for 45 minutes–1 hour. Serve in portions spooned between large macaroons. Pour over warm sauce.

MAKES 6–8 PORTIONS

Tipsy Syrup Tart

100 g (4 oz) fresh breadcrumbs
225 g ($\frac{1}{2}$ lb) golden syrup
30 ml (2 tbsp) dark rum, or grated rind and juice of 1 lemon
100 g (4 oz) butter or block margarine
225 g ($\frac{1}{2}$ lb) plain flour
1 egg, separated
caster sugar

1 Put the breadcrumbs in a large bowl, add syrup and rum or lemon rind and juice and stir until mixed.
2 Rub the fat into the flour until it resembles fine breadcrumbs. Mix the egg yolk with 30 ml (2 tbsp) water and use it to bind the pastry mixture to a firm dough.
3 Roll out three-quarters of the pastry and use to line a 21.5-cm ($8\frac{1}{2}$-inch) loose-based fluted flan tin. Spoon in the syrup mixture and lattice with remaining pastry. Bake at 180°C (350°F) mark 4 for about 20 minutes, until just set but not browned.
4 Brush the pastry lattice with beaten egg white and sprinkle with caster sugar. Return to the oven for a further 20–25 minutes, until well browned. Ease out of the flan tin.

To freeze
Cool, then pack in rigid container and freeze. When required, put the tart back into the flan tin and reheat from frozen at 180°C (350°F) mark 4 loosely covered with a sheet of foil, for about 50 minutes.

MAKES 6 PORTIONS

Butterscotch Ice Cream

90 ml (6 tbsp) soft brown sugar
50 g (2 oz) butter
300 ml ($\frac{1}{2}$ pint) warm milk
2 eggs
65 g ($2\frac{1}{2}$ oz) sugar
5 ml (1 tsp) vanilla flavouring
284 ml ($\frac{1}{2}$ pint) whipping cream
shredded almonds, toasted

1 Warm the brown sugar and butter together in a saucepan until both have melted; bubble for 1 minute. Add the warm milk, heat gently until evenly blended, stirring.
2 Beat the eggs and sugar together well, then stir in the warm milk and vanilla flavouring. Strain back into the pan. Stir over a low heat until the custard thickens slightly. Do not boil. Cool.
3 Lightly whip the cream and mix into the custard. Pour into a container to a depth of at least 5 cm (2 inches) and freeze until mushy. Beat, then return to the freezer until firm. Transfer to the refrigerator 45 minutes–1 hour before serving.

To freeze
Treat as ice cream (see page 61). Serve topped with toasted almonds.

MAKES 8 PORTIONS

Almond and Coffee Meringue Cake

3 egg whites
175 g (6 oz) sugar
50 g (2 oz) ground almonds
50 g (2 oz) unblanched almonds
50 g (2 oz) caster sugar
284 ml ($\frac{1}{2}$ pint) double or whipping cream
45 ml (3 tbsp) coffee-flavoured liqueur

1 Draw two 20.5-cm (8-inch) circles on non-stick paper and place them face down on baking sheets.
2 Stiffly beat the egg whites then gradually whisk in the 175 g (6 oz) sugar keeping the mixture stiff, then fold in the ground almonds.
3 Spread out the meringues into the circles marked on the paper. Bake in the oven at 150°C (300°F) mark 2 for about 1 hour until browned and crisp. Cool slightly, then ease off the paper and place on a wire rack to cool completely.
4 Meanwhile, place the whole almonds and caster sugar in a small pan. Heat gently until the sugar melts and caramelises, turning the nuts over occasionally. When the sugar is well browned pour the nut mixture out on to an oiled work surface and leave to cool and harden.
5 Ease the cold praline off the surface and remove any surface oil with absorbent kitchen paper. Grind through a nut mouli or crush in a pestle and mortar.
6 Whip the cream until it holds its shape. Whisk the liqueur and three-quarters of the praline into half the cream and use to sandwich the rounds. Pipe rest of cream on top and decorate with reserved praline.

To freeze
Pack in a rigid container and freeze. To use, while still frozen, lift out and place on a serving plate. Thaw in the refrigerator for about 2 hours.

MAKES 6–8 PORTIONS

INDEX